Women, Doctors and Cosmetic Surgery

Women, Doctors and Cosmetic Surgery

Negotiating the 'Normal' Body

Rhian Parker
Associate Professor, The Australian National University

First published 2010 by
PALGRAVE MACMILLAN

Palgrave Macmillan in the UK is an imprint of Macmillan Publishers Limited, registered in England, company number 785998, of Houndmills, Basingstoke, Hampshire RG21 6XS.

Palgrave Macmillan in the US is a division of St Martin's Press LLC, 175 Fifth Avenue, New York, NY 10010.

Palgrave Macmillan is the global academic imprint of the above companies and has companies and representatives throughout the world.

Palgrave® and Macmillan® are registered trademarks in the United States, the United Kingdom, Europe and other countries

ISBN-13: 978-0-230-57400-7   hardback

This book is printed on paper suitable for recycling and made from fully managed and sustained forest sources. Logging, pulping and manufacturing processes are expected to conform to the environmental regulations of the country of origin.

A catalogue record for this book is available from the British Library.

A catalog record for this book is available from the Library of Congress.

10 9 8 7 6 5 4 3 2 1
19 18 17 16 15 14 13 12 11 10

Printed and bound in Great Britain by
CPI Antony Rowe, Chippenham and Eastbourne

To all my family, and particularly the women, who are all great role models.

Contents

Acknowledgements

I would first like to acknowledge the assistance, time and willingness to share their experiences that participants gave this study. Without such generosity research such as this would not be possible.

I am grateful for the support of a number of colleagues and friends for their advice, guidance and patient encouragement over the course of this project. I would particularly like to acknowledge the contribution of Paul Komesaroff, Leon Piterman and Kelsey Hegarty.

I thank my family, Stephen, Ruth, Hannah and Alice for accompanying me on this journey and for never doubting my ability to complete the book. Above all, I thank my parents who are no longer with me but who taught me about the value of learning and instilled in me the confidence to achieve.

1
Introduction

Everyone who opens this book will have a view about cosmetic surgery and anyone who watches television or reads women's magazines can not fail to notice the growth in the number of programs and articles dedicated to cosmetic surgery. Whatever we think about cosmetic surgery, it is becoming increasingly common and accessible. While some evidence suggests that men are becoming increasingly interested in cosmetic surgery, it is women who are the main recipients of cosmetic surgery and of the media's attention. Because of this, this book only deals with women and cosmetic surgery and it unashamedly situates cosmetic surgery as a gendered practice. That is, women are over-represented as cosmetic surgery patients and men are over-represented as cosmetic surgery practitioners. This book tells of the cosmetic surgery experiences of both women and their doctors. It describes an empirical study that seeks to unpick these experiences and make sense of the process of cosmetic surgery. It does not pretend to be a theoretical précis, but it does critique past theorising in this area and attempts to move the theoretical debate beyond the duality of women as victims and women as agents towards an understanding of the complex interactions within cosmetic surgery and the pivotal role the doctor plays in the outcomes of cosmetic surgery.

Academics have detailed the extent of the media's obsession with cosmetic surgery. For example, Brooks (Brooks, 2004: 42), in a study of women's magazines in the US, found a large number of articles on cosmetic surgery in four prominent and popular magazines. Fraser (Fraser, 2003b) in an Australian study discusses the way cosmetic surgery is covered in women's magazines and the way in which the advertising of cosmetic procedures became more obvious during the 1990s. Women are the main consumers of cosmetic surgery and are the main participants in media stories that focus on cosmetic surgery. There are a number of TV 'makeover' programs beamed around the world that enthral us by what doctors can do to transform women's faces and, indeed, their whole bodies. We are captivated by the extent of the transformation without really understanding the pain and potential risk these women undertake to be reinvented. We are also enthralled by how doctors seem to magically use their powers to make any woman 'beautiful'. There is a silent juggernaut that is slowly but surely claiming a significant place in our culture, becoming so normal in magazines that we are surprised when there is not an article about cosmetic surgery and taking its place regularly on television programs that expose cosmetic surgery disasters and applaud cosmetic surgery triumphs. Although, in the past people were only exposed to the cosmetic surgical interventions film stars and TV personalities chose to undergo, cosmetic surgery is now within the grasp of 'ordinary' women and these women are accessing it at an increasing rate. Medical practitioners are also publicising cosmetic surgery as accessible and affordable and advertising a range of cosmetic procedures in newspapers and glossy magazines that offer women the opportunity to change the way they look and the way they feel about themselves.

Given that cosmetic surgery saturates the media, why is it that we know very little about what women say about their experiences of cosmetic surgery, why they want cosmetic surgery and the risks they are prepared to take to have cosmetic surgery done? Why is it that we understand very little about what happens between women and their doctors when they engage with each other in the cosmetic surgery clinic and know even less about why doctors practise cosmetic surgery and how they describe how they interact with women who come to them seeking cosmetic surgery?

For the purposes of this book, cosmetic surgery is defined as any medical intervention that is aesthetic rather than reconstructive. It applies to procedures that use surgical and non-surgical tools to change the way a woman looks. These include procedures such as a facelift, brow lift, rhinoplasty or breast implants, as well as procedures using laser and other technological tools. Women who had only received collagen or Botox treatments were not included in the study as these procedures offer short-term benefits and are now regularly carried out by a range of practitioners who may not be medically qualified. While the list of available procedures is constantly growing and the lines between beauty therapy and medical procedures are sometimes blurred, the procedures undergone by the women in this study were those that are only available through medical practitioners.

Cosmetic surgery is about physical appearance rather than physical function. Cosmetic surgery can change the body so as to conceal the impact of time on the body and the impact of bodily functions (such as pregnancy and breastfeeding) on women, or it can remould a physical feature. Cosmetic surgery also enables women to change their bodies to comply more closely with the culturally and socially defined visions of women in Western society.

It is a controversial practice that engenders lay debate and some academic discussion and sits somewhat uncomfortably at the periphery of conventional medicine. Academic comment has variably seen women accessing cosmetic surgery, on the one hand, in terms of agency and self-determination (Davis, 1995), as taking control over an errant body, or, on the other hand, as culture imprinting itself on women's bodies (Morgan, 1991; Wolf, 1990). Gilman (Gilman, 1999) argues that cosmetic surgery is about 'passing' in society, about fitting in rather than standing out. To date, the emphasis of such debates has been on the motivations of, and pressures on, those who choose to undergo cosmetic procedures. The role of the medical practitioner in the process of cosmetic surgery and the nature of the interaction between doctor and patient in this setting has not been looked at or critiqued.

There are conflicting positions within feminist theory about women's rationales and incentives for having cosmetic surgery. Gillespie (Gillespie, 1996: 77) describes these positions as a 'paradox of choice'. Thus while both sides approach the issue from a feminist perspective, they are deeply conflicted. On the one hand, there are those who see women who have cosmetic surgery as victims to cultural oppression (Bordo, 1993; Wolf, 1990). On the other, Davis (Davis, 1993, 1995) argues that relegating women to 'cultural dopes' does not allow for women's self-determination or individual agency. Gillespie (Gillespie, 1996: 82) acknowledges that women's choices to have cosmetic surgery can be seen as rational decisions that may improve their social influence and economic status. She identifies that

> [c]onformity to social and cultural norms may, on the one hand, represent collusion by women in dominant constructions of femininity, nevertheless, at the individual level it may also be rational and empowering for the woman.

Thus, oversimplifying and polarising women's individual experiences of, and motivation for, cosmetic surgery ignores the complexity of women's relationship with their own bodies and also ignores the role of the surgeon in sculpting women's bodies through surgery.

Those who access cosmetic surgery are often reluctant to publicise the fact. They are keenly aware of the societal ambivalence regarding the practice. Even though cosmetic surgery is routinely discussed in the media, the discussion is often in the context of outstanding success or awful disaster. Thus, it is a practice that titillates the media while being shrouded from any real critical attention or debate.

Nevertheless, cosmetic surgery has become a booming industry in many countries across the Western world. Although there are no central data available in Australia on the number of procedures carried out each year, estimates in the Cosmetic Surgery Report (Health Complaints Commission, 1999: 7) suggest that '50,000 cosmetic surgical procedures … and about 250,000 cosmetic medical procedures' were undertaken in 1998. The report also says that the industry doubled between 1993–8. If this is the case, then it can be expected to continue to grow. Nearly 11.7 million cosmetic surgical and non-surgical procedures were performed in the United States in 2007, compared with 6.9 million in 2002 (American Society of Aesthetic Plastic Surgeons, 2008). According to the American Society of Aesthetic Plastic Surgeons, males had 9 per cent of all cosmetic procedures, while females had 91 per cent. The percentage of procedures attributable to males and females respectively was virtually unchanged from 2001. This underlines the fact that it is a particularly gender-specific practice with anywhere between 80 per cent and 100 per cent of patients (depending the procedure) being women. The British Association of Aesthetic Plastic Surgeons (*The Times*, 29 January 2005) reported that cosmetic procedures in the United Kingdom rose by 50 per cent in 2004. For women, the most popular procedures were breast implants, facelifts and eye surgery.

In Australia, a range of medical practitioners carry out cosmetic surgical and non-surgical procedures. The *Cosmetic Surgery Report* estimated that 37 per cent of procedures were carried out by plastic surgeons, 23 per cent by cosmetic physicians, 13 per cent by dermatologists, 11 per cent by ear, nose and throat and other specialists, 10 per cent by nurses and 6 per cent by cosmetic surgeons. The report also found that there were 350 doctors who had a 'substantial practice' in cosmetic surgical procedures and about 150 doctors and 50 nurses who provide cosmetic medical procedures. Cosmetic surgery consumers reported that liposuction and breast augmentation were the most popular procedures. Multiple procedures of the face, facelifts, dermabrasion, brow lift etc. were also popular, as was rhinoplasty. Over 60 per cent of consumers surveyed were between the ages of 15–44 and only 20 per cent of respondents were on an income above $A60,000 per annum. Unsurprisingly 86 per cent of those who had cosmetic surgery were women.

Why study cosmetic surgery?

Although significant numbers of women choose to undergo cosmetic surgical and medical procedures, little is known about their experiences of this process or of their interaction with the doctors who operate upon them (Fraser, 2003a). Doctors have been criticised by feminist writers for 'colonising' women's bodies by marketing their services specifically to women and making cosmetic surgery appear as normal for women as a 'hairdo' or wearing make-up (Dull and West, 1991). However, there is little research on how doctors and patients interact in this setting or on what influence, if any, the doctor has on the outcome of cosmetic surgery. Are they merely technicians who carry out procedures or do they bring to bear into the process their own attitudes and understandings about women?

What is specific and different about cosmetic surgery is that its aim is not just to change physical features but to change the vision women have of themselves. The perplexity of cosmetic surgery is that what is at stake for women is the very constitution of the self. Through surgically altering their bodies, women seek to redefine their identity and their sense of who they are and how they appear to the world. They need to change a part of their body that they feel is at odds with that sense of self. The aim is to achieve a physical transformation that better reflects who they feel they are and how they project themselves to the world around them. To do this they must engage with the doctor. Women and doctors need to achieve a shared appreciation of what it is that women want from cosmetic surgery and what doctors can provide. This exercise of negotiating the outcome of surgery is at the heart of the practice of cosmetic surgery, and this exercise comprises translations of meanings and communication across diverse discursive boundaries. The only way that this exercise can be untangled and understood is by probing the nature of the communication process and interaction between doctor and patient in this setting and plotting the course of each of their experiences and the meanings behind them. By doing this, points of commonality and difference can be identified and isolated. Once this is understood, then the aetiology of potentially discordant discourses can be explored as can their effects on the outcomes of cosmetic surgery.

There is an impossible tension in cosmetic surgery. The patient base is predominately women and the doctors are predominately men.

This must inevitably contribute to inconsistent, contradictory and divergent assumptions about the nature of women's bodies and what constitutes female beauty and physical acceptability. As neither women nor doctors live in a cultural and social vacuum, their experiences of gender and physical appearance must be both socially and culturally influenced. Do these influences contribute to the discourses surrounding cosmetic surgery and to its outcomes? At stake for women who seek cosmetic surgery is their sense of self and personhood, but the only way they can redefine their sense of self is by interacting with doctors and the medical paradigm. This book outlines the distinct and divergent discursive practices that women and doctors engage in during the process of cosmetic surgery. These discordant discourses are central to the process of cosmetic surgery. Until they are understood and the gap between them negotiated, there will be, and indeed *should* be, continuing concern about the appropriateness of cosmetic alteration as it is currently practised.

In order to understand these discourses, and what other factors influence the outcomes of cosmetic surgery, this book will chart the structure and dynamics of the interactions that constitutes a woman's journey through the process of cosmetic surgery. To do this it utilises a qualitative technique where the individual narratives of both women and doctors are recorded and explored. Consistent and discrepant discourses and descriptions are mapped and set in a phenomenological description of the women's and doctors' journey through cosmetic surgery. This enables points of contact and distance between the two to be identified.

Overview of the book

This book is about the repertoire of influences that impact upon the process and outcome of cosmetic surgery. It argues that women who seek cosmetic surgery have a vision of how they want their bodies to look. This vision is constructed through their subjective experiences of the world, gendered experiences and cultural notions of what constitutes femininity in Western societies. The limited empirical evidence we have available (Davis, 1995) suggests that women's motivations for having cosmetic surgery are based on the wish to fit in to society and be 'normal'. As any perusal of women's magazines would show, the 'normal' is often far from the 'real' vision of women's bodies. Thus, as women read these texts and view images of other women's bodies from such publications, they are exposed to an idealised and airbrushed view of the female body. While most women would never expect to look like these magazine models, they do provide a contrast against which to measure their own bodies and impact on their body images. The literature about how body image issues impact on women is reviewed in Chapter 3.

If women are affected by social and cultural attitudes regarding the female body and how these are reflected in the media, then men's visions of women must be impacted too. Such visions must have some influence on how male doctors who carry out cosmetic surgery see women's bodies and on how they interpret what women want from cosmetic surgery.

If women and their doctors bring with them to the process of cosmetic surgery a vision of women influenced by current cultural stereotypes, does this influence the outcomes of cosmetic surgery? This book argues that such views, attitudes and opinions percolate through language and discourse in the negotiation about the nature and extent of the cosmetic alteration to be undertaken and, ultimately, to the outcomes of cosmetic surgery.

In remaking themselves through cosmetic surgery women may be seen to be exercising agency (Davis, 1995) or submitting to a culture that has a specific and idealised notion of female form and beauty (Morgan, 1991). They may be making an informed choice to emulate that form or they may be acceding to unrelenting cultural and media-mediated pressures. Regardless of what motivates women to seek cosmetic procedures, the outcome of that procedure in not in their hands alone; it is very much an outcome negotiated between them and their doctors.

Medical intervention at the site of the female body is not a new phenomenon. In reproduction and other 'natural' functions medicine has actively intervened. Illich (Illich, 1976) noted that human experience is increasingly coming under the scrutiny of medical practitioners, what he saw as the 'medicalization of life'. Morgan argues that Western medicine has traditionally seen women's bodies as sites of intervention, as 'generally inferior, deformed, imperfect, and/or infantile' (Morgan, 1991: 157). Women's bodies are sites to be worked upon and normal female bodily functions something to be medicalised and controlled. Given this, it is not surprising that cosmetic surgery would be the next medical intervention on women's bodies. Balsamo (Balsamo, 1996: 208) asserts that

the cosmetic surgeon's gaze doesn't simply medicalize the female body, it actually redefines it as an object for technological reconstruction.

Medical practitioners are also schooled in a profession of pre-eminence and power where their expertise and medical knowledge carries with it authority. The combination of authority and their own culturally defined attitudes regarding the female form makes it difficult for them to be merely neutral participants in the act of changing the way a woman looks.

What women end up with after a cosmetic procedure is a result that is affected by a plethora of discourses and meanings, mediated by techno-logical (medical) knowledge/power, historical and cultural definitions of femininity and the normative definitions of ideals of feminine beauty. The resultant interplay between these factors is the substance of this book.

Outline of the study

This study consists of interviews with 32 women who had undergone a cosmetic procedure and 19 medical practitioners who carry out these procedures. Of the women interviewed, 9 had formally complained to the Health Services Commissioner of Victoria about their treatment. These women were included in the study so as to understand what had led them to make such a complaint. All the interviews were conducted in Victoria, Australia. Pseudonyms have been given to all the women and the doctors quoted in this book. The interviews focused on three components that were postulated to be integral to understanding the process of cosmetic surgery. They are viewed from both the doctors' and from the patients' perspective. These are:

- The motivation of patients and doctors. Why women undergo cos-metic surgery, why doctors carry it out and how each group reflects on each other's motivations,
- Communication between doctors and their patients. What consti-tutes communication and how is this communication achieved, and
- The perception and understanding of what constitutes risk to women and doctors in this setting.

The influences these components have on the outcomes of cosmetic surgery are varied and complex and the data on this are set out in Chapters 5, 6, 7 and 8.

Background discipline of doctors

Doctors who carry out cosmetic surgery as a significant part of their medical practice come from a range of medical disciplines. No specialist training is mandated before offering aesthetic procedures to the public. In this study, 14 of the doctors interviewed were plastic surgeons, four were dermatologists and one was a cosmetic surgeon. Plastic surgeons undergo specialised training through the Royal Australasian College of Surgeons (RACS) and are fellows of that college. Dermatologists undergo training in dermatology and become Fellows of the Australian College of Dermatology. Cosmetic surgeons do not need to undergo any specific training through any college.

Gender of cosmetic surgeon

Four of the women interviewed had been operated on by female doctors and three of these four women had been operated on by the same female doctor. Given that women are under-represented in all areas of specialist medicine, this is not surprising. Women mentioned that one of the reasons for choosing a female doctor was that they would understand what women wanted:

> You see I thought that she would know what a woman wanted.
>
> (Dora)

However, this was not borne out in practice as it seems that there was little attempt on the part of the female doctors approached by women in this study to really explore women's needs or wishes.

Number and type of procedures

The number of non-surgical procedures carried out by each of the doctors interviewed ranged between 100 and 7000. Those that carried out the most procedures practised cosmetic medicine as well as cosmetic surgery. This includes laser resurfacing of the face, Botox and collagen injections, chemical peels, fat injections to the lips and liposuction. Dermatologists carried out more cosmetic medical procedures than the plastic or cosmetic surgeons. Some of these procedures, particularly Botox and collagen injections and chemical peels, were carried out by nurses in the practice.

Plastic and cosmetic surgeons carried out between 100 and 1000 surgical procedures a year. They were less involved in medical procedures and more involved with surgical procedures. The surgical procedures they offered were facial surgery, breast surgery and liposuction[1]. Dermatologists were more involved in non-surgical procedures, such as laser resurfacing, although some did carry out surgical procedures such as facial surgery and liposuction.

How long doctors had been practising cosmetic surgery

The doctors interviewed had been carrying out cosmetic surgery for between 4 years and 35 years. Over half the doctors had been practising cosmetic surgery for between 10 and 20 years and four doctors had been in the industry for over 20 years.

Procedures women reported having undergone

The procedures women reported having were breast implants, breast reduction, liposuction, facelifts, brow lifts, eye surgery (blepharoplasty), cheek implants, nose surgery (rhinoplasty) and laser resurfacing. Of the 32 women interviewed, five had undergone cosmetic surgery on more than one occasion. Other women had undergone more than one procedure but these were as part of the same operation, for instance, a facelift and a brow lift. Both the women who had undergone breast reduction surgery had done so some years ago and their experience of cosmetic surgery explored for this book related to their most recent experience of surgery.

Characteristics of women interviewed

Women in this study came from a range of backgrounds and professions and were aged from the mid-twenties to the late fifties. All the women were either in paid employment or were retired from full-time paid work. None of the women went in for cosmetic surgery on a whim. Many had thought about it for a long time. Those who had undergone liposuction had tried diet and exercise before seeking cosmetic surgery. What was common for all the women was the extent to which a particular feature of their body bothered them. It was a constant negative presence in their lives and affected the way they saw themselves and related to others. While they were mindful of being judged by others for their decision to have cosmetic surgery, they were prepared to accept such criticism rather than continue to put up with their body as it was. Theirs was a lonely journey, with very few women mentioning that they had been supported by friends or family when they went to appointments with their doctor. It was also a secretive journey, as there was a palpable reluctance among the women to share their 'secret' with anyone but close family or friends.

This study does not purport to delve into the individual psychological reasons that may lead women to undergo cosmetic surgery nor does it look at psychological conditions, such as body dysmorphic disorder. This has been done elsewhere in a number of studies (Cash, 1990; Cash et al., 2002; Pruzinsky and Edgerton, 1990; Sarwer et al., 1998). It instead focuses on the social nature of the female body and how social discourses impact on self-imagery and identity.

Chapters in the book

This book consists of 10 chapters that are organised into three sections. These sections are: literature reviews (Chapters 2, 3 and 4), a presentation of the results of the empirical study (Chapters 5, 6, 7 and 8) and a discussion of the main themes of the book and a conclusion (Chapters 10 and 11).

The literature review covers three domains that contribute to an understanding of the story of cosmetic surgery and of how women's bodies are presented and perused in Western society.

Chapter 2 is the first chapter of the literature review. Its purpose is to review the available literature that outlines a history of cosmetic surgery. It provides part of the theoretical framework that explains the role of the doctor in cosmetic surgery.

Chapter 3 discusses literature on women and body image. There is a wide range of literature in this area and this chapter provides a brief overview of that which is most pertinent to women and cosmetic surgery.

Chapter 4 takes a wider perspective and discusses literature from philosophy, feminism and sociology so as to situate cosmetic surgery in the literature on the body from these disciplines.

Chapter 5 explains how cosmetic surgery is very much a commercial endeavour where a battle for business between practitioners is part of the subtext of the industry.

Chapter 6 deals with the first of the key themes, the motivations behind women's decisions to undergo cosmetic surgery, the motivations of the doctors who practise it and how each group understands each other's motivations.

Chapter 7 deals with communication in the cosmetic surgery setting. It explores the role and nature of communication in the cosmetic surgery clinic from both the patients' and practitioners' perspective.

Chapter 8 looks at the understanding of patients and doctors about what constitutes risk in the cosmetic surgery setting. It also reports on what risks are detailed to women and how these risks are explained to them.

Chapter 9 provides an overview of the findings outlined in Chapters 5–8. It details how the outcome of cosmetic surgery for women is achieved through a complex and dynamic repertoire that is the result of interrelated factors which are often at odds with each other. It also pulls together some of the existing literature in the area and offers a critique of the debates in this literature moving the argument forward against the background of the experiences of the women and doctors that were part of this study.

Chapter 10 draws together the main themes and arguments of the book and suggests that we need a more sophisticated and multi-disciplinary analysis of cosmetic surgery. Cosmetic surgery is here to stay and more attention needs to be paid to improving the quality of the cosmetic surgery process, to accountability and to the outcomes of surgery. At the very least we need to know how much cosmetic surgery is carried out and how much of it lives up to the expectations of those who choose it as an option.

Cosmetic surgery needs to move from the peripheral shadows of medicine, to be made accountable and to lay itself open to critical scrutiny. It is only then that there will be enough publicly available information for informed choices to be made about cosmetic surgery and about where the power lies in the cosmetic surgery process.

2
Cosmetic Surgery and the 'Normalisation' of the Body: A Short History

Surgery carried out for cosmetic reasons became increasingly popular during the twentieth century and has established itself as a significant, if somewhat controversial, practice in modern medicine. Despite a significant resurgence and growth during the twentieth century, cosmetic surgery is not a purely twentieth-century phenomenon but has a long and complex history. This chapter will sketch what is known of the history of cosmetic surgery and how it has been practised over time in different parts of the world.

Aesthetic or cosmetic procedures are reported to have been undertaken over the past two and a half thousand years. While this chapter cannot begin to comprehensively trace the history of cosmetic surgery across centuries and civilisations, it will provide a brief outline of significant milestones in the evolution of cosmetic surgery as a discipline. Two thorough historical accounts of the development of surgery for aesthetic reasons (Haiken, 1997; Gilman, 1999) have been published and the evidence from these texts, as well as other sources, will be included here. While both these publications focus on the US, Gilman does explore the history of European cosmetic surgery as far back as the Renaissance. Scant attention has been paid to the history of cosmetic surgery in Australia or in the UK. What little material is available is referred to in this chapter.

The first sites of intervention

It is not surprising that the development of cosmetic procedures is interlinked with the development of reconstructive and plastic surgery. The relationship between the two, however, has always been fraught with difficulty, with those identified as plastic surgeons from the beginnings of the specialism being careful to differentiate themselves from 'beauty' doctors. While this tension still remains, many plastic surgeons in Australia, the US and the UK are increasingly devoting their time to cosmetic work rather than confining themselves to purely reconstructive procedures. The demarcation line now seems to be between those trained in plastic and reconstructive medicine and those who identify themselves as cosmetic surgeons. The tensions this creates in contemporary cosmetic surgery will be discussed later in this chapter.

The beginnings of cosmetic surgery are firmly linked to the face, a human being's window to the world and the world's window to that human being. It is here that the discipline began and here it continues to operate.

The nose and cosmetic alteration

The history of surgery carried out for cosmetic reasons can be traced as far back as about 600 BC. The focus of surgery at the time was on the restoration of the nose and earlobes. Procedures to reconstruct noses that had been amputated as a punishment for a crime are reported to have been carried out in India in 800 BC.[1] The incorporation of the Indian and Arabian surgical practice into Western thinking commenced during the time of Aegineta (625–90 AD) in the Roman Empire. Wallace (1982) reports that reconstruction of the nose was carried out near Delhi, India, around 1000 AD. Generations of the same family are said to have performed the operations continuously between 1440 and 1937. The first known written details of these operations were published in India in 1794. The article describes a technique whereby a flap of skin from the forehead is drawn down and used to reconstruct the nose. This technique was known as Indian rhinoplasty.

It was not until the Renaissance period that Vesalius' anatomical text *Fabrica* was published. This was the foundation stone upon which modern surgical practice rests. Gaspare Tagliacozzi, a surgeon in Bologna around mid-sixteenth to late sixteenth century, made the distinction between *chirurgia curtorum per institonem* (surgery healing by grafts) and *chirurgia decoratoria* (beauty surgery). This distinction reappears in the nineteenth century when Ziess published *Handbuch der plastischen Chirurgie* in 1838 (Wallace 1982). However, the actual term 'plastique' was used by Desaut in 1798 and was derived from the Greek *plastikos*, meaning to mould.

During both the seventeenth and nineteenth centuries syphilis was a virulent disease and carried with it significant stigma. Syphilis is caused by bacteria that attack body's mucous membranes and is usually transmitted through sexual contact. Before modern antibiotic treatments were available, one of the most obvious signs of the disease was the syphilitic nose. This was caused by the collapse of the cartilage and an infection of the bone. It was also a disease transmitted from mothers to babies during pregnancy, as a result of which babies could be born with significant deformities.

Those who had the physical symptoms of syphilis were anxious that these manifestations, and the stigma that went with it, be less visible to society and sought to have their noses rebuilt. So as Gilman (1999: 10) says:

Plastic surgery was initially understood as surgery of the nose.

At this time, it was men who were, in the main, the recipients of this surgery. Aesthetic surgeons of the time could make people 'pass' in a society that had become increasingly moralistic, the nineteenth-century Victorian era in Britain being an example of this.

At the same time, the surgeons themselves were 'elbowing' their way into the medical establishment. They were passing from being seen as 'quacks' and occupying a marginalised position in medicine to being part of the medical establishment, even if they were not wholly accepted by it. As they levered to 'pass' into respectability, so they assisted patients to cross the same boundary. Gilman (1999: 22) sees this model of 'passing' as

the most fruitful to use in examining the history and efficacy of aesthetic surgery.

'Passing' in society became an issue again early in the twentieth century after the First World War when many servicemen received terrible facial injuries, particularly in the trenches where the head was most vulnerable. In order to be able to function in a post-war life, many soldiers underwent major facial reconstruction for functional as well as for aesthetic reasons.

Others, before and since that time, have undergone nose surgery, in particular, to enable them to 'pass' in a society that has undercurrents of racial prejudice. Gilman notes that the Jewish nose, the Irish nose and the Oriental nose were sites of alteration. Gilman (1999: 90) says that

[t]he original nose, the normal nose, the healthy nose is that of the European, which may be altered through cultural interventions, but remains a sign of the universality of all human beings.

So as not to stand out in Eurocentric American society, people of Jewish or Asian descent sought to have their noses changed. This has increasingly become the case in multicultural societies such as Australia. Here there is increasing anecdotal evidence that those from Jewish, Asian, Mediterranean and Middle-eastern backgrounds and of both genders seek rhinoplasty.

The face and cosmetic alteration

A resurrection of plastic or aesthetic surgery took place during the 1840s. A German facial surgeon Friedrich Dieffenbach, considered the father of modern plastic surgery, was at the forefront of developments in facial surgery. Dieffenbach differentiated between what he saw as the 'real medical' function of plastic or reconstructive surgery and the merely 'aesthetic' function of some of the procedures carried out. He used the terms aesthetic and cosmetic in a pejorative way. This theme was continued by others in the field. For instance, in 1934 Jaques Maliniak commented on the 'brazen quackery' that he saw was part of plastic surgery (Gilman, 1999: 12). This tension has continued, but now is more focused on the differentiation between plastic and cosmetic surgeons rather than within plastic surgery.

A significant milestone in the history of plastic surgery came during the First World War. During this period surgeons on both sides of the Atlantic were faced with terrible maxillofacial injuries of the war victims. Surgeons worked on both restoring adequate function to the face and on the patient's appearance. The restoration of appearance was seen as imperative if injured soldiers were to be able to earn a living and be accepted back into society. Haiken (1997) notes that equating a man's appearance to his economic status had been established during the nineteenth century and the market value of this appearance was directly linked to his ability to be economically productive. Thus surgeons worked to reconstruct the faces of soldiers so that they would better fit back into society. It was possible to undertake such work due to the development of anaesthesia in the mid-nineteenth century and the increasing adoption of antisepsis in surgery (Gilman, 1999: 16).

The techniques developed during this period led American surgeons, in particular, to return home and practise plastic surgery. The English, however, did not embrace the specialty as wholeheartedly. By the mid-1930s there were still only four plastic surgeons practising in Britain, only one in France, but in America there were at least 60.

The alteration of physical features was now done for functional and aesthetic reasons and the lines between reconstructive surgery and aesthetic surgery were blurred. The ability to change a feature to mask one's ethnic heritage or a disease was now established and accepted and was the prerequisite to contemporary cosmetic surgery practice. Gilman (1999: 16) contends that

> [o]nce you can change what a society understands as unchangeable, such as racial markers, then it is possible to imagine altering other aspects of the body that seem permanent, such as signs of aging. The historical development of specific procedures mirrors the unique double face of aesthetic surgery as parallel to and different from reconstructive surgery.

Establishing the specialism

The development of plastic surgery as a medical discipline occurred at a varied pace in America, Europe and Australia. Not surprisingly, the development of the discipline in America occurred much faster than elsewhere.

Plastic surgery in America

Gilman (1999) reports that as early as the American Civil War, reconstructive surgery was being performed on the faces of soldiers. One of the surgeons, Gurdon Buck, began taking 'before and after' photographs of these soldiers and these were published in 1876. Other American surgeons followed suit and took photographs of patients. The place of plastic surgery was firmly established in America after the First World War, with surgeons returning from the war to establish the specialism as an academic discipline.

The American surgeons were also positioned to take advantage of an increasingly consumer-driven culture, and it was women who were at the forefront of seeking cosmetic alteration during this time. In 1923, the American comedienne Fanny Brice had her nose 'bobbed', an event that was performed as a publicity stunt in a hotel room in Atlantic City. Brice was a well-known and long-time member of the Ziegfield Follies in the 1910s and 1920s. Dorothy Parker commented that 'Fanny Brice cut off her nose to spite her race' (Goldman, 1993). The operation was conducted by a Henry Schireson who later became infamous as a result of numerous lawsuits and scandals surrounding his practice. However, other surgeons were more successful and their efforts brought certain respectability to the industry. They were anxious to differentiate themselves from 'quacks' and began the process of organising themselves into a professional body.

Although Brice's surgery was regarded as a publicity stunt, it did influence the way people saw cosmetic surgery. By the end of the 1930s, the consensus in magazines was that this surgery was a 'miraculous' treatment for inferiority complexes and many other physical conditions deemed to have psychological or psychiatric causes. Underlying much of the media discussion about cosmetic surgery was the release from distress. At that time, doctors who performed cosmetic surgery echoed this theme, seeing themselves as healing the mind by changing the body.

In 1921, the American Association of Plastic Surgeons (AAPS) was founded. This association had as a prerequisite for membership the requirement of degrees in both medicine and dentistry. In 1931, partly to counteract the restrictive rules of the AAPS, the American Society of Plastic and Reconstructive Surgery (ASPRS) was formed. The ASPRS was quick to gain ascendancy and soon became far more influential than the AAPS.

The first meeting of what was to become the American Board of Plastic Surgery took place in 1932, the first annual meeting in 1938 and by 1941 the Board had published its first list of certified plastic surgeons. The journey to this position had been fraught with conflict and difficulty. Tensions were apparent around the fundamental role of plastic surgery. While the First World War established it, in the minds of some, as a reconstructive discipline, others had, before then, seen it as aesthetic in nature. As American and other Western cultures moved away from the moral and social strictures of the nineteenth century, appearance became more important in society. Ideas about physical beauty also changed and surgery became a tool through which beauty could be bought. Here was an opportunity for surgeons to make money, and many took up the challenge. At the same time, a 'fault line' was developing within the group of doctors who carried out reconstructive and plastic surgery.

Even though the Board was established in 1941, the stage had been set for many practitioners to follow the demands of the market and carry out mainly cosmetic work. Those who took the reconstructive path, however, also carried out surgery for cosmetic reasons but engaged with terminology that medicalised certain conditions and reframed them as 'deformities'.

Surgeons who carried out plastic surgery were keen to differentiate themselves from those who carried out 'beauty' surgery, although, like today, the differentiation between the two is often difficult to identify. At the same time, doctors who carried out cosmetic procedures began to see their work as promoting the mental health of their patients by changing their physical appearance. As will be discussed later in this book, this trend has continued. Now, however, the focus has moved from the syphilitic nose or the war-injured face to women's bodies.

Such tensions still remain but have had little real impact on the exponential growth of cosmetic surgical procedures, and the doctors who carry them out. While America led the profession, other Western countries have followed, and the cosmetic surgery industry continues to be lucrative for those practitioners who practise it.

Plastic surgery in the UK and Europe

Other than what is known about plastic surgery in Renaissance Europe and the developments in techniques during the First World War, there is little information available on how the specialism developed in more recent times. Rogers (1976) describes the development of medical texts in cosmetic surgery. Between about 1912 and the 1930s, an increasing number of books and articles were published in America and Europe on cosmetic surgery techniques. He lists how surgeons such as Eugen Hollander (1867–1932), Jacques Joseph (1865–1934) and Erich Lexer (1867–1937) of Berlin describe techniques for face-lifting.

The first known plastic surgery procedure to be performed in the UK was carried out by Joseph Carpue in 1814. Carpue operated on the nose of a military officer. More than a century later, in 1946, the British Association of Plastic Surgeons was founded with 36 people attending its first meeting (Wallace, 1997). In 2006 it changed its name to British Association of Plastic, Reconstructive and Aesthetic Surgeons in response to the changing nature of the work carried out by its members.

In 1926, Dr Suzanne Noel, a French doctor who practised cosmetic surgery, wrote a handbook, *La Chirurgie Esthetique, Son Role Social*. Davis (1999: 478) says that Noel

> was interested in gaining recognition for a new and controversial medical practice as well as perfecting its procedures and techniques.

Davis describes Noel as 'a fervent believer in cosmetic surgery'(ibid.) who viewed cosmetic surgery as 'a social necessity, particularly for women' (ibid. 479). Noel was also a feminist and she belonged to women's organisations that pressured for women to have access to work and the professions:

> She was convinced that cosmetic surgery alleviated suffering and was a useful tool for helping women – to be sure, affluent, professional women – to achieve financial independence and social recognition.
>
> (Davis, 1999: 486)

During this period of the twentieth century in France, and elsewhere, women were becoming increasingly involved in public life and fashion, and cultural ideas of beauty were changing. The hemlines of women's dresses were now below the knee rather than at the ankle, and the fashion demanded that women have flat chests and 'boy-like' figures. Comiskey (2004) describes a surgical procedure carried out on a young woman in France who wanted thinner calves. The surgery was a disaster in that one of her legs had to be amputated. The legal case that followed described how Suzanne Geoffre, a young fashion designer, wanted thinner calves now that French fashion had hemlines on the knee. The outcome of the first judgement in the case stated that doctors did not have the right to operate on a healthy body for the sake of beauty. However, surgeons argued that

> because beauty was so essential in finding a mate and in obtaining and keeping a job, they provided an important service and should be allowed to practise cosmetic surgery.
>
> (Comiskey, 2004: 32)

An appeal upheld the initial judgement but utilised different reasoning. The appeal let the surgeons off the hook as the judges stated that they did not want to denounce the use of all cosmetic surgery but stated that it should be used with 'caution' (Comiskey, 2004).

This case illustrates a number of things. Firstly, it shows a direct link between how the change in fashion and culture led to women accessing cosmetic surgery. Secondly, it demonstrates how doctors were becoming increasingly involved in cosmetic surgery as a means of enabling women to fit in with current fashion and cultural ideals in 1920s France. Finally, it shows that such intervention to abolish 'ugliness'[2] was now unstoppable in spite of the risk involved.

Plastic surgery in Australia and New Zealand

The participation of Australian and New Zealand Army Corps (ANZAC) in the First World War led to ANZAC doctors, together with their British and American counterparts, receiving plastic surgery training. Taylor[3] notes that there were four generations of Australian plastic surgeons. The first generation surgeons were trained during and after the First World War but were not solely plastic surgeons. The second generation trained during, and in the years after, the Second World War mainly in the UK as part of the armed forces.

The third generation, those who trained during the 1950s and 1960s, received the initial part of its training in Australia and New Zealand and then was trained further in the UK. The fourth generation, those who are currently practising, usually trained at home and then went to the UK or the US for further training.

Taylor notes that after the Second World War there were seven plastic surgeons in Australia and five in New Zealand. By 1956, when the Royal College of Surgeons established a Plastic and Reconstructive Surgery section, there were 17 founder members from Australia and four from New Zealand.

The increase in the public profile of cosmetic surgery in more recent years in Australia as a beauty tool paralleled that in America. Once it came to be openly discussed in the media, particularly in women's magazines and on health-related television programmes, the phenomenon was bound to grow. The rise in interest and publicity coincided, to some extent, with the ability of doctors to advertise their services. Until the early 1990s doctors in Australia were prohibited from advertising. When this changed, it enabled practitioners to publicly compete with each other for business. In cosmetic surgery, this meant that there was now a different marketplace. No longer were women only given referrals by their general practitioners or a discreet recommendation to a particular doctor, the doctors were now able to openly tout for business. Thus, instead of cosmetic surgery being seen as something that Hollywood film stars indulged in, or as the pastime of the rich and famous, it was now portrayed as accessible to the average person. Advertisements suggested that women could get what *they* wanted out of cosmetic surgery. The increasing impact and availability of cosmetic surgery meant that doctors in Australia were now not only treating injuries or curing diseases, but were instrumental in shaping the physical forms of women and making them blend into a society where beauty has such a high premium.

Conclusion

While the discipline of cosmetic surgery continued to develop, a nagging problem worried many early practitioners. If medicine was meant to heal, what role did aesthetic surgery have in medical practice? Some plastic surgeons in the early years of the discipline in America scorned cosmetic surgery and claimed it went, 'against the medical profession's fundamental principles' (Haiken 1997). However, the pace of change rapidly taking place in Western society meant that cosmetic surgery was gaining patients and practitioners. At the same time psychology was gaining ground as an academic discipline and was beginning to be understood by lay people. It is not surprising then that the next step was to see cosmetic surgery as a tool for improving a person's self-image, by treating physical 'defects' or 'abnormalities'. Those who practised cosmetic surgery saw their discipline as providing a physical solution to psychological problems. Haiken (1997: 95) points to surgeons being

captive, as well as captains, of the new enthusiasm for psychology

This new approach meant that by the early 1940s most plastic surgeons in America performed some cosmetic surgery as part of their work.

As the psychological arguments for cosmetic surgery became embedded in the discipline of plastic surgery, so did the notion that appearance was vital to success. This idea had been promulgated since at least the 1920s and was now gaining momentum. At the same time, women's magazines were becoming increasingly popular and were an ideal avenue for selling goods and services to women. In such publications natural phenomena, such as ageing (specifically for women), were being problematised. Dealing with wrinkles and the sagging face through the facelift was offered as a simple solution to an age-old 'problem'. Magazines gave singularly positive write-ups on all things to do with cosmetic surgery. The range of procedures available has, of course, grown and there is now the ability to inject chemicals into the face to erase frown or smile lines and to 'plump up' the face. For plastic and cosmetic surgeons around the world, this ability to 'improve' psychological health has been the linchpin argument for the promotion of the practice of cosmetic alteration.

It may be useful here to return to Gilman's notion of 'passing'. His thesis that aesthetic surgery is fundamentally about enabling people to 'pass' in society is supported in contemporary Australia by the way cosmetic surgery is discussed in the media and advertised by practitioners (Fraser, 2003b). As we are increasingly kept under surveillance, both by ourselves and others, and the expectations we have of our bodies become greater, cosmetic alteration seems to be an easily accessible solution to perceived bodily deficiencies. We may no longer see the effects of syphilis on the noses of sufferers, nor the terrible damage of trench warfare on the faces of soldiers, but the need to 'pass' in a society that pays such great attention to looks is just as real.

Cosmetic surgery developed as a response to the need to 'pass' or blend into society and practitioners constructed noses and faces that would enable this to happen. The medical industry that historically developed around the way the body looks reflects the intricate interaction between medicine and culture. The practice of cosmetic surgery could not survive nor grow without the cultural imperatives and pressures that drive us to change the way we look. With the increased exposure of women, and their bodies, in twentieth-century society, the cultural lens became increasingly focused on how women looked. There was a constant weight of cultural and social expectations on women to look different or 'better'. In response to this, cosmetic surgery has thrived and those who practise it sculpture women's bodies in line with social and cultural expectations of how women should look, so that they can 'pass' in a society that scrutinises women and their bodies closely.

In order for their work to 'pass' as medically respectable in their own minds, and in order to explain it to their peers, those who carry out cosmetic surgery continue to justify their work as psychologically necessary and to reframe many natural processes as 'deformities' that need treatment. However, they tend not to acknowledge the social and cultural basis of their work. This theme has a presence across the history of cosmetic surgery and it will be explored further in this book.

If cosmetic surgery is about 'passing' in society, those who perform it are conduits for prevailing cultural norms and expectations. Thus, at a time when large breasts and small hips are promoted as attractive, those who carry out cosmetic procedures respond by providing women with these features. Similarly, when youth is applauded and the signs of ageing are seen as flaws, practitioners respond by making women look younger. All these procedures that react to cultural demands on women also reflect the way surgeons see women's bodies and what they can provide to women on the operating table. While there may be negotiation between doctor and patient about a procedure and its outcome, the ultimate power is in the hands of the party that holds the scalpel. This has been true throughout the history of cosmetic alteration. However, the way cosmetic procedures are marketed in contemporary society promises women a transformation that accords with their view of themselves and their needs. It promotes the notion that women can take control of their bodies. Nowhere does it suggest that this control may be in the hands of the surgeons who operate on them. In the past, recipients of cosmetic surgery may have been grateful for any improvement in the way they looked. However, as cosmetic surgery has become increasingly available and promises so much more, recipients expect to be given what they have asked for and not what the surgeon wants to, or is able to, provide.

3
The Female Body and Body Image: A Historical Perspective

This chapter describes historical and contemporary ideals of the female body and discusses the way in which body image has become a particularly significant issue for women in contemporary Western societies. It looks at how women's size and the way they look have become so important to them and so debated in the media and across academic disciplines. Although there is some evidence to suggest that young men, and in particular gay men, are becoming more body conscious, the 'unbearable weight' (Bordo, 1993) of bodily surveillance and body image concerns is still very much a female burden.

The female body in history

Notions of beauty are not constant. They are historically specific and culturally constructed. The term 'beauty' in contemporary Western culture is almost always used to describe the female rather than the male body. In Western societies today, women's body image is an issue that is much discussed in the media and by women themselves. It forms the backbone of a large international weight loss and cosmetic surgery industry. The same could not be said for men. As Hesse-Biber (Hesse-Biber, 1996: 20) says:

> Our culture judges a man primarily in terms of how powerful, ambitious, aggressive and dominant he is in the worlds of thought and action. A woman, on the other hand, is judged almost entirely in terms of her appearance, her attractiveness to men, and her ability to keep the species going.

Women's ideal body image has depended upon standards of beauty and desirability at different periods in history in different cultures.

For instance, Akhenaten's queen *Nefertiti*, translated as 'the beautiful woman has come' is one of the most well known of the queens of Egypt. She was known at the time for her outstanding beauty. Her bust was discovered in 1912 and has since become one of the best-known images in history and viewed as the ideal of the female face.

Artists across the centuries have painted women who were considered to be beautiful during their era. Renoir's famous painting, *The Bathers*, painted in 1887 depicts two naked young women with voluptuous thighs and hips and cellulite! Compared with current ideas of beauty, where women are much slimmer but have bigger breasts, this voluptuousness is almost shocking. Catwalk models are now almost all painfully thin, as are most female celebrities, and these are the contemporary visions of women we have somehow come to expect and accept.

Historically the size and attributes expected of women's bodies were dictated, in the main, by male desire and marriageability. In Neo-Confucian China for more than a 1000 years, the feet of higher-class girls were bound, the bones being broken to create a small lotus-like clubfoot:

> The bound foot, a symbol of feminine beauty, represented a woman's only prospects in life ... Unable to inherit property or pass on an ancestral name, a girl was an economic liability until she left to join her husband's family ... Her only value lay in her marriageability.
>
> (Hesse-Biber, 1996: 21)

In sixteenth- and seventeenth-century Europe, a small waist became fashionable. In fact Queen Catherine of France introduced the idea of binding the waist so that the circumference could be minimised to the ideal thirteen inches (Baker, 1984). During the seventeenth century, while the waist was still laced, breasts became more fashionable and women with ample breasts and buttocks were celebrated. A pale pallor was also fashionable and white lead-based paint was used to achieve this effect.

Tight corsets reappeared in Europe and North America during the mid-nineteenth century and were a compulsory accessory for the middle-class woman. The rise of Western capitalism meant that industrial work was separated from the home and the wives of middle-class businessmen could join those of the upper-class landed gentry in being 'a prized showpiece of her husband's wealth' (Hesse-Biber, 1996: 23) A small waist was a prerequisite for these women and:

> symbolised passivity, dependence, and, more perversely, bondage ... To attain such an ideal, a tightly laced undergarment reinforced

with whalebone, and later steel, constricted women's waists for many hours a day. This pressure often caused pain and distorted the internal organs and rib cage.

(Hesse-Biber, 1996: 24)

In an era when marriage was vitally important to women, particularly for economic security and status, conformity to the accepted body image norm was very strong. Unfortunately, these tight corsets were not only painful to wear, but could cause miscarriages, fainting and internal organ damage.

While in the nineteenth century the thin waist was achieved through the corset, ample hips, as shown in the paintings of artists such as Renoir, were also seen to be desirable. At the turn of the twentieth century, the fashion demanded a small waist, pigeon-breasted bust and full 'swayback' hips. The early twentieth century brought greater political emancipation for women, but also encouraged the culture of thinness. Newly emancipated young women of the 1920s were not constrained by the corsetry of their mothers or grandmothers, but instead, by the amount they ate. In order to achieve the slender boyish bodies that were fashionable at the time, they starved themselves and bound their breasts.

Western countries during the twentieth century increasingly promoted thinness, even emaciation, as the ideal for women, while in the past larger women were thought to be beautiful. In Greek and Roman times, representations of women were often muscular and rounded. Even as recently as just over a century ago, the ideal woman's body was quite the reverse of what it is today. Seid (1994: 5) says:

the female ideal was Junoesque: tall, full-busted, full-figured, mature. Dimpled flesh – what we today shudderingly call 'cellulite' – was considered desirable. Sinewy, 'close to the bone' women 'no bigger than a whipping post' suffered disdain, not those with amply fleshed curves properly distributed and disciplined by the corset. The undergarment industry even came to the aid of the slighted thin woman with inflatable rubber garments (replete with dimples) for her back, calves, shoulders and hips.

Women who were well rounded were regarded as healthy, with a good temperament and as displaying 'disciplined habits' (Seid, 1994: 5). It was at the turn of the twentieth century that things began to change. Seid observes that the transformation happened 'for a variety of reasons' including the increasing place of 'speed and motion' in society' (Seid, 1994: ibid.). Despite the slimming down of the ideal form, plumpness was still regarded as a sign of health and well being. This situation remained fairly constant through the difficult years of the Depression in Western countries and throughout the Second World War, where food was rationed in many countries. Women's limited access to food, and their increased involvement in manual work during this time, meant that they were slimmer by circumstance rather than necessarily by choice.

After the Second World War, during the 1940s and 1950s, women were back in the home and embracing the girdle rather than the corset. The ideal body images of the time were fuller, with an 'hour-glass' figure (achieved with the girdle). This could have been a reflection of increased access to food and increasing affluence in Western countries. Women's magazines emphasised women's role as wives, mothers and homemakers and fashions were geared to complement these roles.

In the 1960s, the Western world was faced with a number of revolutions: the rise of feminism, the availability of the contraceptive pill, colour television, the peace movement and a growth in the representation of women in the workforce, in particular, the professions. The pressure on women to compete in a male-dominated workplace and to dress to the fashion of the time again dictated that they should be thin. The most famous fashion model of the time was Twiggy and young women strove to emulate her thin, wide-eyed look.

Women's magazines reinforced the 'thin' ideal, with slim waif-like models again defining the trend. Thus, while women, and young women in particular, were receiving messages about liberation, equality and emancipation, they were also receiving messages that indicated that all of this happens only to *thin* women.

Body image and fitting in to contemporary society

Orbach (1999: 1) says that

> [t]he twentieth century, particularly the latter part of it has created a notion of the body that we find hard to inhabit.

The body is no longer central to the maintenance of daily living. Increasingly, activities once carried out by the body are now conducted by machines. Add to this the rise in capitalism and consumption and what we now have is a body whose identity is linked less to the work it does and more to the way it looks. Orbach goes on to claim that

> we have become enslaved not just to consumerism but to the body as our personal product which we must shape and reshape according to the dictates created in the market but felt individually.
>
> (Orbach, 1999: 2)

and:

> Globalism, while promising variety, the exotic and the unknown, evacuates the richness of difference. It imposes an intensity of imagery around beauty, youth, health, consuming, that negates the possibility of the very kind of potential pluralism implicit in cultures connecting with each other. Where social intercourse across continents suggests and promises melange, the commercial exigencies of globalisation attempt to control plurality and to decrease self expression ... In place of multiple images of beauty ... we get reduced images of beauty. We get lean Asian women, lean African women, lean Indian women ... Similar bodies allowed to be multicoloured or multi faced but no different than that.
>
> (Orbach, 1999: 3)

Such uniformity of look and shape means that women all over the world fret about their shape and the way they look. Thus, Asian women seek out surgery to make their eyes and the bridge of their nose look more Western. Kaw (1994) interviewed Asian-American women who had undergone cosmetic surgery of the eyes and/or nose to give them a less-Asian look. She reports that these women felt constrained by their Asian features and that 'with the critical eye of the oppressor' (1994: 260) had internalised negative images of Asian-Americans in a 'postmodern culture (that) actually obscures differences' (ibid.).

Similarly, Jewish and Middle Eastern women want to have smaller, less distinctive, noses. While it can be argued that women who have these procedures just want to 'fit in' to the society they live in, there is little doubt that the way women's bodies are portrayed allows for little difference in size, shape or particular ethnic features. Women's bodies need to conform to those idealised images that are so prevalent around us.

Across history and civilizations, bodies have been fashioned to conform to standards of the time. As Sullivan (2001: 2) observes:

> The diversity of body customs has led anthropologists to conclude that a body is both a physical and symbolic artefact, forged by nature and by culture at a particular moment in history. Social institutions, ideology, values, beliefs, and technology transform a physical body into a social body. The resulting social body bears the imprint of the more powerful elements of its cultural context. Bodies, therefore, provide important clues to the mechanics of society.

In contemporary Western culture there is ample evidence of the use of the body to communicate group membership and social identity. Tattooing the body has been popular among some men to symbolise belonging to motorcycle gangs. Body piercing and tattooing has become increasingly popular among adolescents as a symbol of fashion, a means of differentiation from previous generations and prevailing norms, and as a sign of belonging to a particular group. Of course, these adolescents may not see their piercing and tattooing as signs of fitting in, but rather as resistance, of standing out. Bordo (1993) maintains that such body customs reflect cultural oppression rather than empowerment. Thus, the decision a young woman makes to pierce and tattoo her body is less a sign of agency than a means of fitting into a particular social milieu that itself is heavily embedded within a social and cultural context.

Body image and contemporary Western women

While the standardised Western body that we see daily in magazines and in the electronic media has become the norm in idealised appearance, medicine and the health industry have worked in partnership to advise and cajole us to eat in a certain way. While the commercial availability of 'fast foods' has eased the burden of food preparation for women, it has also introduced us to larger portions and higher calories. As an antidote to this, weight has become a medical issue and guidelines as to what we should eat to remain slim and healthy have become entrenched in the discourse of health promotion. With an increasing emphasis on 'healthy eating' and on thinness as a sign of health and vitality, it is little surprise to see the effects of these messages and of the messages that promote idealised visions of the female form, on contemporary Western women.

There is now increasing evidence to show that women, from a very early age, are affected by images they see in the media. Research shows that the media has a stronger influence on the body image of adolescent girls than adolescent boys (McCabe and Ricciardelli, 1999). Other research (Paxton and Durkin, 1999) has found that some girls, having viewed advertising images of thin young women, are more susceptible to body image disturbance than others who have viewed them. What we do not yet know is why this is so. An interesting study comparing the anthropometric characteristics of models, shop mannequins and Barbie found that

> [a] young woman randomly chosen from the reference population would have a 7% chance of being as ectomorphic as a catwalk model, a 3% chance of matching an international model, a 0.3% chance of matching a shop mannequin, a 0.1% chance of matching a 'supermodel', and no chance at all of matching Barbie.
>
> (Olds and Norton, 1999)

The fact that since 1959 Barbie has been on the list of best-selling toys and that 95 per cent of girls in the US aged between 3 years and 11 years own at least one Barbie 'makes her clearly a force to contend with'(Urla and Swedlund, 2000: 398)

Barbie:

> remains an incredibly resilient visual and tactile model of femininity for prepubescent girls headed straight for the twenty-first century.
>
> (Urla and Swedlund, 2000: ibid.)

Not only did Barbie exemplify the perfect female form, she also brought with her a whole range of accessories. In the new era of post-war affluence, young girls became consumers of Barbie fashion and were introduced to the new world of consumer capitalism.

There is little doubt that the images of women that we have been presented with over the past four decades are out of the reach of most Western women. This leads, inevitably, to young women, and indeed, not-so-young women striving to achieve the impossible through unhealthy weight-loss strategies and, increasingly, surgical intervention. There is a body of evidence now available that shows most women in Western cultures to be dissatisfied with their bodies. Evidence from America (Lamb et al., 1993; Wardle et al., 1993), Britain (Wardle et al., 1993) and Australia (Huon et al., 1990) indicates that when shown silhouettes of female figures, women rated their figures both as substantially larger than the ideal figure and what they thought would be most attractive to men. Thus, it is not that women necessarily want to look like Elle MacPherson, Cindy Crawford or Gemma Ward, but this is the image of contemporary beauty projected, and this image includes large breasts and very narrow hips, full pouting lips and not a blemish in sight. These are the images that women see around them on billboards, in magazines, on television and film every day. It is, therefore, hardly surprising that some of the messages that these images project have a significant impact on women in Western, and increasingly non-Western, societies. There is intense social pressure on women to be attractive and this pressure is not confined to young women but affects women of all ages. Such pressure is not static but with increasing demands upon women to be competitive in the workforce as well as raising a family, it is constantly evolving. Such an evolution makes cosmetic surgery an attractive solution to those who feel their bodies do not match contemporary standards.

Research carried out on the ideals of breast and chest size in the 1990s (Tantleff-Dunn, 2001), for instance, found that the ideal breast size preferred by men was larger than that preferred by women and, significantly, breast size and its association with a range of positive characteristics was markedly larger in 1998 than it was in 1992. Why this is so is not clear. The implications of this for women and their dissatisfaction with their bodies, and in particular their breasts, may be to turn to surgery in increasing numbers so as to conform to these expectations. Here they may be putting themselves in the hands of male doctors whose preference for larger breasts may impact on the outcome of breast augmentation surgery.

Why are women more concerned than men about their bodies?

Given that the pursuit of thinness pervades our culture and, increasingly, images of taut and muscular men are being displayed on billboards, why is it that women seem to be more affected than men by these messages? Firstly, in contemporary Western culture, women's bodies have been open to the public gaze for a long time. Females have been paraded in beauty pageants and their bodies used to sell a vast array of products and services. Biologically women's bodies find it hard to fit into the current ideals with their hips failing to obey the standard set by models and the effects of reproduction confounding the ideal of the board-like abdomen.

In practical terms, and in order to compete in the workplace, women's economic survival is too closely linked to beauty and, as was discussed earlier in this chapter, the ability to attract the opposite sex. Despite the strides forward in equality of opportunity achieved by women, the rules of contemporary Western culture still perpetuate values and norms that require certain defined standards of behaviour and attractiveness in women. While men, and young men in particular, may worry about weight and muscle mass, they are not subject to the stringent bodily control that surrounds women. Young men are expected to be unconstrained in sport and social life, while young women are expected to exhibit restraint. This control extends to diet and body size and to the constant battle against calories.

While women are considered to have been 'liberated' over the past three or more decades, social surveillance over the female body has increased. As a woman's body ages, she is no longer acclaimed for her experience and achievements but watched for signs of disintegration. She is constantly reminded that, at a cost, she can access a vast array of creams and lotions that can stave off wrinkles and sagging and any other sign of decay. Her task is to remain trim and taut despite the years. Men on the other hand, as they age, are thought of as having achieved status and experience. Grey hair and sagging jowls are not an impediment to work promotion and a man's identity is not tied up to how his body presents itself. The visual ideals of men found in the media are far more varied than those of women. Male power and influence is not dependent on his body, but on his wealth and social status. Ageing does not impair pop idols like members of the Rolling Stones from continuing to perform, however 'worn' Keith Richards may look. In fact they are applauded for their longevity in an industry where young women need looks more than talent to survive. The visual diversity accepted for men allows them to feel comfortable in their bodies despite age and size.

An exception to this can be found in gay men, who score consistently higher in measures of body dissatisfaction than heterosexual men (Lakkis and Ricciardelli, 1999). This dissatisfaction is often displayed through disordered eating, bulimia and dieting and is attributed to a homosexual subculture that highlights a slender and muscular body ideal. While some research has emphasised sexual orientation as the most significant causal factor in body dissatisfaction and the related behaviours, other studies point to gender-related personality traits as having the key influence. Thus, gay men with negative femininity traits, that is, traits such as dependency and timidity, were more likely to have disordered eating. While no research findings are yet available on the extent to which gay men access cosmetic procedures, anecdotal evidence suggests that they are doing so at an increasing rate and this may be because of the pressure exerted from within their culture to achieve a 'perfect' body.

Women and the control of their bodies

Given the overwhelming pressure on women and their bodies, can women remain in control of a body that is under constant social surveillance? Is it possible that a woman can mould and shape her body to achieve acceptability in her own eyes and in the eyes of a culture that sets such strict parameters? Bartky's (1990: 42) view is that such control is not within women's grasp:

> Woman has lost control of the production of her own image, lost control to those whose production of those images is neither innocent nor benevolent, but obedient to imperatives that are both capitalist and phallocentric. In sum, women experience a twofold alienation in the production of our own persons: the beings we are to be are merely bodily beings: nor can we control the shape and nature these bodies are to take.

Findings from the Australian Longitudinal Study on Women's Health (Kenardy et al., 2001) indicate that amongst a sample of 14,800 women aged 18–22:

- 47.8 per cent had dieted to lose weight within the last year,
- 12.2 per cent had dieted five or more times within the last year,
- 67.5 per cent of the women who were overweight dieted to lose weight in the last year and
- 20.9 per cent of underweight women were also dieting to lose weight.

The study also found that

> Both weight and shape dissatisfaction were significantly greater with higher dieting frequencies.

One of the most disturbing findings for this population was the mean onset age for dieting of 15.4 years. Ironically while women are far more likely to be dieting than men, men are more likely to be overweight. The Australian Institute of Health and Welfare reports that the prevalence of obesity rose significantly during the 1990s, by 71 per cent for men and 80 per cent for women (Australian Institute of Health and Welfare, 2003). Current figures suggest that 16 per cent of men and 17 per cent of women over the age of 18 are obese and a further 42 per cent of men and 25 per cent of women are overweight but not obese.

Women's attempts to control their bodies through diet and exercise regimes are constant and, as mentioned earlier, begin at a young age. The advertising industry is central to promoting products and practices to women that encourage them to take control over their bodies by eating specific products, following specific diet plans and signing up to exercise routines. However, Wolf (1990) argues that rather than empowering women, such routines are controlling and make women passive. The fit and lean body does not change the power women have in society; it merely makes them conform to culturally and socially constructed restrictions.

Like dieting, the lure for women who undergo cosmetic surgery for rejuvenation or alteration is to look better, thinner, to achieve an improved self. The medicalisation of beauty has spawned an industry that promises to recreate a woman in her own image, but advertises using images of women that fit only a small proportion of the female population. While saying it can achieve what women want, the industry ignores the influence social and cultural standards of beauty have upon it and on the way it perceives women. While it is dependent on women's dissatisfaction with their bodies, the industry also creates a market by tantalising women with 'easily achievable' beauty. Of course the reality of cosmetic surgery is the same reality as any other medical intervention. It has real risks, depends on the doctor's technical skill and his/her ability to interpret patient wishes. The process of cosmetic surgery is complex and dynamic and the interaction of factors such as understanding patient and doctor motivation, effective communication and the way patient and doctor see risk, will have a greater or lesser effect on the outcome of any procedure. Added to this is the influence the doctor's own vision of women has on the process. Given all of these influences, women who decide to undergo a cosmetic procedure may well be risking losing control of their bodies to a medical intervention whose outcome is by no means certain. Their notion of 'beauty' and the 'acceptable body' may well not mirror that of the doctors who will change their bodies forever.

Conclusion

Gender differences are not only biologically determined, culturally constructed, or politically imposed, but also ways of living in a body and thus of being in the world.

(Blake, 2000: 430)

Women across centuries have lived in bodies that beauty standards of their times have moulded and delineated. While it was expected that women have large hips and thighs at the turn of the twentieth century, they were expected to be very slim a mere 20 years later. When a woman's economic survival was tied to marriage, she was expected to bind her feet or whalebone her waist to catch a suitable partner. While the physical consequences of such actions were painful and crippling, it was the price women paid to be socially acceptable. It is little different in contemporary Western culture where women are pressured to conform to certain standards of weight and attractiveness. Those who become fat, or whose breasts are too small, noses too large or wrinkles too noticeable, judge themselves through self-surveillance and the surveillance of others, including that of other women.

Blake (2000: 458–9) argues that foot-binding in neo-Confucian China was a voluntary act carried out by mothers on their daughters so that they could succeed in a world controlled by men:

> The rhetoric of foot-binding ... connected human effort and individual initiative on the part of women to the ontology of femininity, to the idea of self-control over individual fates, to the social roles that allowed women to fulfil their purpose as bearers of sons, to the fertility of the earth, and to the cosmic order of things. It was powerful mystification of gender because it entailed the mindful bodies of women overcoming intense, protracted physical pain in a drama that coupled 'self-sacrifice' with 'self-exaltation'.

Extreme diet and fitness regimes and cosmetic surgery are contemporary equivalents of foot-binding. The coupling of sacrifice and pain to enable women to conform to contemporary demands on their bodies.

But what of those women who cannot or will not conform to cultural expectations? Those who are overweight or obese are seen as lacking self-control, as lazy, as self-indulgent. Obesity is also often seen as avoidable for the wealthy and as a condition of the lower socio-economic classes, with obesity among women of lower socio-economic classes in the US being seven times more prevalent than among those in the highest socio-economic class (Fallon, 1994).

As cosmetic surgery becomes more common, those women who have the economic means will be able to access it, but those without those means will not. This has the potential to create further division and discrimination for women anxious about the way their body looks. This will also be the case for those women who choose not to have cosmetic surgery.

Ironically while women suffer increasing body dissatisfaction, obesity rates are rising rapidly (Paquette and Raine, 2004). Women are getting contradictory messages: on the one hand, thinness is promoted as the ideal and on the other, the consumption of the "wrong" types of food is encouraged by the food industry. Women turn to dieting and unhealthy weight-loss regimes, and when these fail, cosmetic surgery can provide a cure through liposuction. Paquette and Raine (ibid.) caution that the role of health professionals in controlling women's weight can contribute to body image dissatisfaction and support the socio-cultural environment that does not allow for a wider range of healthy body sizes.

Women who undergo cosmetic surgery will be following historical precedent and adapting their bodies to better conform to prevailing standards of beauty and acceptability. Deciding to have cosmetic surgery may be a woman's individual choice and an empowering decision, but it is made within a particular social and cultural context. Making this choice also extends medicine's role in disciplining the body and enables medicine to play an active part in the aesthetics of the body so that it reflects and perpetuates socially and culturally acceptable standards of beauty for women.

4
Social and Feminist Theories and the Body

This chapter sets a range of theoretical approaches to the body and its place in the social world. These theoretical approaches help situate the practice of cosmetic surgery in contemporary Western society. As we have seen in the previous chapter, body image is a pervasive presence in women's lives. The way women interact with their bodies and experience them is socially and biologically determined. While no one theoretical approach can fully explain or encapsulate cosmetic surgery, between them they provide a platform that enables us to question some assumptions that are the common parlance of cosmetic surgery. For instance, some feminist theorists characterise cosmetic surgery as simply a repressive or disciplinary regime (Morgan, 1991; Wolf, 1990). Others (Davis, 1995) see women's choice of cosmetic surgery as a matter of exercising 'agency' over their own body and how it looks. Fraser (2003b) sees this repertoire of agency as being reflected in magazines which often characterise women as actively taking charge of their bodies through their decision to undergo cosmetic surgery.

Whether women are 'dupes' or 'heroes', these characterisations focus only on their choices about undergoing cosmetic surgery. They do not extend to the reality of the process and outcomes of that surgery.

Social and feminist theories about the body, particularly the female body, provide a backdrop against which a theoretical agenda about cosmetic surgery can be addressed. These theoretical paradigms also contribute to an understanding of the discursive positions that are an inherent and problematic part of the cosmetic surgery process. While women make choices about their bodies, those choices are acted upon by doctors who bring into the process their own set of discourses and understandings about the female body. These two clusters of discourses are informed by the embodied experience of both doctor and patient. In turn, those experiences are social, psychological and culturally constructed. It is through the lens of these disparate influences that the theatre of cosmetic surgery is carried out.

Social theory and the female body

There is a range of theoretical paradigms that can contribute to an understanding of the practice of cosmetic alteration of the body. Philosophical thinkers since ancient Greece have discussed the role of the body, its relationship to reason, the mind and the soul. Plato saw the body as a type of prison wherein lay the mind, reason and the soul. The body was a battleground where reason should rule over the appetites of the body and the soul. Christianity developed this distinction between matter and form, between the soul that was given by God and the body that was always susceptible to sin and carnality. Developing from these traditions, philosophical thought saw consciousness as separate from the body.

The work of more recent philosophers has contributed much to a rethinking of the way we see the body. For instance, Merleau-Ponty's phenomenology of the body identifies the lived experience of each body and takes it beyond its purely biological function. Merleau-Ponty (Bigwood, 1991; Merleau-Ponty, 1992) saw that:

> [m]y body does not appear to me as an object, a set of qualities and characteristics to be linked up with one another and thus understood … I and it form a common cause, and in a sense I am my body. Between it and me there cannot be properly said to be a relation, since this term designates the behavior of one object in reference to another.
>
> (Merleau-Ponty, 1992: 108)

He saw the body as inexorably intertwined with its environment and the images reflected from that environment onto, and about, the body as part of the embodied subject. Interpreting Merleau-Ponty's phenomenology of the body, Bigwood (1991: 61) notes that his 'body-subject':

> has a unique sensitivity to its environment … The body is actively and continually in touch with its surroundings.

Thus, it is our body that enables us to have a 'world' and our biological self is entwined in our cultural existence and vice-versa. Merleau-Ponty's notion of embodiment contributes significantly to an understanding of how bodies exist as wholes rather than constituent parts. The body is the coordinator that brings together environment and behaviour. Diprose (1994: 119), in discussing Merleau-Ponty's work, points out that:

> the capacities and habits, and therefore the interests, of any body do not arise separately from its engagement with others nor from the discourses and practices which make up the world in which it dwells ... your corporeal schema is never individual: it is fundamentally inter-subjective and specific to your social and familial situation.

The body, then, is sensitive to its environment and exists:

> simultaneously in cultural and natural ways that are inextricably tangled.
>
> (ibid.: 109)

Merleau-Ponty has, however, been criticised for portraying the body as unassigned in terms of gender, race, and age and, as such, does not take into account the influences these factors have.[1] For women, these issues are crucial to the way they feel about their body and how their bodies are perceived by others.

Foucault's work has also had a significant impact on theoretical conceptualisations of the body. Foucault (1972, 1973, 1979) focused on the disciplinary regimes directed against the body to produce 'docile bodies' that can be controlled to efficiently produce and conform. Foucault (1979) argued that the rise of political liberty and parliamentary institutions in contemporary Western society was accompanied by a new discipline focused on the body. This discipline regulates the body so that it 'produces subjected and practised bodies, "docile" bodies' (ibid.: 138). Foucault utilises the school, the army and the prison as examples of the control and surveillance inflicted upon the body. While Foucault's analysis is theoretically detailed and historically situated, he does not differentiate the body as gendered and does not allow for gendered disciplinary practices. Foucault's critique is thorough and compelling but it has been criticised for ignoring the gendered nature of the body (Bartky, 1998; McNay, 1991).

Culture transmits messages in a gendered way and displays women as squarely in the gaze of others through advertising and other media images. For women, these messages are particularly focused on the way their body appears to, and engages with, the world. Body size, shape and form are regarded as signals to the inner person. It also contributes here to the understanding of the practice of cosmetic surgery and the ways in which we see ourselves as reflected in the eyes of others. Lingis (1996: 65) claims:

> Our 'body-image' is not an image formed in the privacy of our own imagination: its visible, tangible and audible shape is held in the gaze and touch of others ... And I see my own visibility with their eyes, feel my own tangibility with their hands, hear my voice with their ears.

Elizabeth Grosz (1987) has argued that what is needed is a 'corporeal feminism' and has criticised philosophical approaches to the body because, she says, they see the body as one element which includes mind and body, but subverts the body to a mere vessel. Grosz contends that the body should be looked at both from within, that is, psychoanalytically, and from the outside, the way the body is 'lived or experienced by the subject' (ibid.: 9). Here, the body can be conceptualised as the intersection of mind and culture, a place where the internal mind is expressed and the exterior body is the 'inscriptive surface' (ibid.: 10). Thus, as Eagleton[2] states, 'It is not quite true that I have a body, and not quite true that I am one either'. The body is a physical representation of the culture, influences, attitudes and beliefs that we are all susceptible to, but react to in different ways.

Heyes argues that feminists need to adopt a 'richer ethical grammar and vocabulary' (2007: 90) when theorising cometic surgery. She contends that rejecting cosmetic surgery outright as a patriarchal practice and assuming that the cosmetic surgery industry was 'so transparently oppressive' (ibid.: 91) that it required only scant analysis, means that feminists have not fully engaged in the debate about the industry. Jones (2008) outlines the feminist approaches to cosmetic surgery so far and concludes that new approaches are called for given that cosmetic surgery is now well established across western culture ad society.

Grosz (1994: 187), while recognising the contribution of 'key male theorists' who have:

> Individually and collectively ... affirmed that the body is a pliable entity whose determinate form is provided not simply by biology but through the interaction of modes of psychical and physical inscription and the provision of a set of limiting biological codes.

argues that bodies have been presented as neutral and universal, ignoring 'sexual specificity' (ibid.: 188).

Women who have a cosmetic procedure do so against a background of seeing themselves in relation to *their* world while being subjected to the gaze of others, particularly that of the doctor. Diprose (1994: 124) observes that

> biomedicine is not just one among many fields of knowledge which regulate bodies in the name of the so-called common good: it holds a privileged place in disseminating knowledge about what a body is, how it functions and the nature of its capabilities. And, in this, biomedical knowledge does its own dichotomising in delineating the normal body from the abnormal.

Those who practise cosmetic surgery hold such a 'privileged place', as they define what are normal, and abnormal, features of women's bodies. Added to this is the tendency for biomedical science to 'identify women's bodies against a male norm ... [that] ... get reproduced within this scientific knowledge of the body' (ibid.: 128). In cosmetic surgery, doctors have the opportunity to inscribe on to women's bodies the culturally and socially accepted standards of female beauty and form and, in a predominantly male-dominated industry, this means 'male' standards.

While philosophers had long been debating the body, social theorists were fairly late in finding it. Shilling (1994) sees the body as being 'something of an "absent presence" in sociology' (ibid.: 19). He identifies 'four major social factors' (ibid.: 139) as providing the context for the appearance of the body as a subject of sociological inquiry. These factors are:

> the growth of 'second wave' feminism: demographic changes which have focused attention on the needs of the elderly in Western societies: the rise of consumer culture linked to the changing structure of modern capitalism: and a growing crisis in the knowledge of what our body is.

The body has become central to sociological enquiry over the past decade or so. The influence of feminism, the work of Turner (1992: 19) among others, has encouraged a focus on the body as a central part of the human experience and of human action. Bryan Turner claims that 'we can conceive of the body as a potentiality which is elaborated by culture and developed in social relations'. He draws a parallel between the theories of Foucault and those of one of the founding fathers of sociology, Max Weber:

> Although they used a different vocabulary, Weber's interests in how 'personality' is produced has a relationship to Foucault's ideas about the 'techniques of the self'.

> (Ibid.)

Indeed Foucault, together with the influence of feminist thought, encouraged sociologists to look at how the body reflects and is produced by social relations. Pierre Bourdieu *does* see the body as socially differentiated, displaying its social position and culture. Shilling (1994: 130) summarises Bourdieu's perspective in this way:

> bodies are unfinished entities which are formed through their participation in social life and become imprinted with the marks of social class. Bodies develop through an interrelation between an individual's social location, habitus and taste. These factors serve to naturalize and perpetuate the different relationships that social groups have towards their bodies, and are central to the choices people make in all spheres of social life.

The unfinished nature of bodies, according to Bourdieu, enables bodies to evolve and adapt to the social world. In the modern world, the body, the way it presents itself and is cultivated, can add value and prestige to the 'self' and create self-identity. The body in the early twenty-first century is physically malleable and is imbued with value according to the social and cultural standards about the body that prevail. Thus, in the early twenty-first century, the successful body is seen as slender and taut, controlled and cared for and imprinted with the values of contemporary society. As has already been seen in Chapter 3, these values change over time and across societies, but are constant in their influence on the body.

Bourdieu's analysis of the body also includes the process of capital accumulation. Physical capital, the way the body presents itself and looks, influences other forms of capital such as wealth and status. Thus, particular ethnic features may add or remove capital value (Gilman, 1999). Similarly, overt signs of ageing can do the same thing for women, in particular. Women who seek cosmetic surgery to improve their chances of staying in the workplace as they get older can be seen as acting to improve their capital value.

Bryan Turner (1996: 124) discusses the way in which

> Social success depends on an ability to manage the self by the adoption of appropriate interpersonal skills and success hinges crucially on the presentation of an acceptable image.

This acceptable image renders obesity, for example, as 'irrational' and marks it as a signal of an unsuccessful body, one that does not fit well into the disciplined space that we inhabit. Similarly, the ageing body, or the body marked by childbirth and the general wear and tear of life, signals a slippage in the hierarchy of a culture that values the 'look' in an increasingly consumer and consumption-orientated market. On the social stage, the body must now perform and look the part. He argues that the secularisation of society and the 'calculative hedonism required by mass production' (ibid.: 172) has meant that traditional uses of cosmetics and cosmetic adornment and the symbolism that these held, have now been superseded.

> Cosmetic practises are indicative of a new presentational self in a society where the self is no longer lodged in formal roles but has to be validated through a competitive public space.
>
> (Ibid.)

Together with this exposure and need for validation in public, Western society is now able, and indeed often expects to, modify the body through exercise, surgery, cosmetics and dress. Of course, this is nothing new as different tribes and social groups at different times throughout history have inscribed and adorned their bodies. In contemporary Western society, however, rather than this inscription demonstrating traditional signs and adhering to certain rituals, it acts as a means of changing the individual and their identity to fit more closely into their idea of current culturally dominant standards of an acceptable body.

Giddens (1991) suggests that the individual has become more central in late modernity, and taking control of oneself and one's life is an expectation of modern social life. He theorises that in contemporary Western society the individual is responsible for the way they construct their life and what they make of it. In order to do this, each individual is involved in a process of 'reflexively organised life-planning' (ibid.: 5). He also suggests that adapting the self to meet changing demands of relationships with others has become important in a system where relationships are more temporal. As individuals we now move in wider circles and have more liberty than at any time in history and, through the body, we display ourselves on a wider stage. In order to display ourselves to others as we want others to see us, we are no longer tied to our own physiological make up. We can now change our bodies, through the bodily regimes of diet, exercise and surgery, to reflect the way we want to be seen by others. We can exercise agency and self-determination in responding to the demands of modern social life, we can construct our identity, we can overcome fixed bodily features and inherited bodily traits and make the body a site of renovation, a reflexive project that can, through human action, overcome biology and physiology.

Exercising such agency in the decisions we make about changes to our body is, however, mediated by the other actors in the process. In cosmetic surgery the significant 'other' is the medical practitioner who carries out the procedure. Fundamental to the ability to obtain bodily change or reformation through cosmetic surgery is the medical profession and the discourse that surrounds it. The way they interpret our wishes about our bodies, and the way they inscribe their own culturally and socially constructed ideas, has an impact on the outcome of any surgical procedure to change the way we look.

Medical authority and the normalisation of the female body

Sociological analyses of women's health since the early 1970s have pointed to the increasing medicalisation of various aspects of normal female bodily functions (Lupton, 1994; Oakley, 1980; Zola, 1972). The focus has been on women's reproductive functions. Here there has been both surveillance and active intervention. Reissman (1998: 57) contends that

> [m]edicalization has resulted in the construction of medical meanings of normal functions in women-experiences the typical woman goes through, such as menstruation, reproduction, childbirth, and menopause. By contrast, routine experiences that are uniquely male remain largely unstudied by medical science and, consequently, are rarely treated by physicians as potentially pathological.

Surveillance of women's bodies has evolved with the increased availability of technological tools and scientific screening techniques. Thus, sexually active women are expected to undergo regular screening for cervical cancer. Pregnant women undergo repeated ultrasound examinations throughout their pregnancy. Oakley (1998: 138–9) argues that cervical screening is presented as 'a moral duty' for women. It is a practice that is:

> sexualised as an investigation by men (partners and doctors) of women's moral state: positive smears are feared by women for the

implicit stigma of sexual immorality as well as for the implications of serious disease.

Thus, it acts as a surveillance mechanism not only of a woman's reproductive anatomy, but also of her sexual activity.

Active intervention in reproductive functions includes the medicalisation of childbirth and of menopause. Childbirth was traditionally the domain of women as midwives and mothers. Medicine has, however, successfully appropriated childbearing and women's reproductive function has been defined as a 'clinical crisis' (Doyal, 1995). Whether this has led to any health benefit for women is disputed. Cahill (2000), for instance, argues that the decline in maternal and infant mortality rates during the first two decades of the twentieth century in the UK was not a direct result of improved medical care. Rather, it was linked to better diet and standards of living improvement for poorer people. Statistics show that mortality rates fell most significantly during the First World War when 60 per cent of medical practitioners were drafted into the armed forces.

Whatever health outcomes flowed from the medicalisation of reproductive functions, women were brought very much into the medical 'gaze' of doctors and their bodies were defined by medical vocabulary and medical discourse. The normal bodily functions and experiences of women are more likely to be identified and treated medically than those of men (Riessman, 1998). Given that men are overrepresented in the medical profession, this medical scrutiny is most likely to be carried out through male eyes. In this way, both professional power and patriarchal power impact upon women's bodies and their lives.

Riessman (1998: 59) claims that the medicalisation of bodily functions is:

> a contradictory reality for women ... As women have tried to free themselves from the control that biological processes have had over their lives, they simultaneously have strengthened the control of a biomedical view of their experience.

If we accept that medicine has taken possession of women's bodily functions, it is not surprising that it should also get involved in women's aesthetic form. In the same way that ownership and production of resources sits mainly with men so does the ownership and production of scientific knowledge. That this knowledge should be put into use to create more tangible wealth for men from women's bodies is one of the drivers of medicine's involvement in cosmetic surgery.

Riessman (1998: 47) argues that doctors medicalise women's lives and their experiences 'because of their specific beliefs and economic interests'. She contends that they respond to market conditions that serve 'their ideological and material motives'. She also argues that women are complicit in this 'medicalisation process because of their own needs and motives, which in turn grow out of the class-specific nature of their subordination' (ibid.). In this way 'a consensus develops that a particular human problem will be understood in clinical terms' (ibid.). However, this accord is problematic as women can benefit or lose from it. While 'there had been a "fit" between medicine's interest in expanding its jurisdiction and the need of women to have their experience acknowledged' (Riessman, 1998: 57) the cost for women has been to surrender autonomy over their bodies and succumb to medical direction.

At a broader social and economic level, business has also benefited from this expansion of the medical domain as drug companies, pathology laboratories and medical equipment suppliers profit from what happens in the medical clinic. In cosmetic surgery, profit for doctors and suppliers of goods and services are significant and these groups work together to generate new markets.

As will be seen in Chapter 6, women are under no illusions as to doctors' motivations for carrying out cosmetic surgery. They see doctors' involvement in the area as economically driven. In the same way that other female bodily functions are disciplined by medical intervention, so cosmetic surgery acts to regulate how a woman's body will look.

Oinas (1998: 54) states that

> [m]edicine is not an objective science in the sense that it exists outside, or above, culture and society. On the contrary, medicine is part of culture, a social institution that is both formed by society and an active agent forming that society.

Like any branch of medicine, cosmetic surgery functions within culture and society, it reflects its ideals and, at the same time, helps create them. While women cannot be seen as willing supplicants to medical imperialism over the body, they do give their body over to be altered by that medical intervention.

During the 1990s there was a shift from the biomedical model of surveillance to a population based or psycho-social model (Nettleton, 1996). The health narrative now shifted to self-monitoring and self-control over the body to avoid certain risks through adopting a healthy lifestyle.

Such a shift coincided with, or even generated, the broader attention to the body as a social rather than a primarily personal space.

This model of health has continued to encourage women to think beyond the health of the inner body to focus on the external aesthetic body. Thus, while women may be independent agents who make individual choices about cosmetic surgery, those choices are impacted upon by social and cultural ideals about the body and health. These ideals are impregnated into the process and discourse of cosmetic surgery not only by the visions *women* have of how they should look to be healthy, but also by the expert knowledge and the socially and culturally constructed ideas *doctors* bring with them.

If doctors as experts do bring with them such ideas into their practice of cosmetic surgery then through what mechanism are such ideas translated onto women's bodies? One framework that can help explain this is that offered through the work of Michel Foucault (1972). Foucault introduced the notion of discourse into philosophical, cultural and sociological thought. One interpretation of his concept of discourse sees it as being situated in, and produced at, the time that it is created. At certain points in time we think about, see, discuss and write about certain things in particular ways. Thus, what we see is affected and modified by historical, social and political influences. At the same time, discourse also shapes meaning and provides a framework for understanding the world around us. However, there is no common discourse shared by everyone. For instance, there can be a competing set of discourses that are influenced by power, authority and practice. Thus:

> There are a number of discourses operating in our society at any one time, and at times these discourses find themselves in competition with each other as they offer different ways of conceiving of, and explaining, the world. Thus, in the health care arena we have present at any one time many different discourses varying in their authority and influence.
>
> (Cheek et al., 1996: 174)

Foucault (1973) suggested that medical and scientific discourse dominates Western society to such an extent that other discourses with different frameworks and possibilities are excluded.

Foucault (1973) discusses the medical and scientific pursuit to 'normalise' the body. The definition of 'normal' is based on the discourse of influential groups, for example doctors, who define what is 'normal' in body shape, size, behaviour and health. Thus, a regime of diet, exercise and, increasingly, surgical intervention, can make bodies 'normal' and more acceptable to Western notions of beauty. Foucault also sees that individuals are under the constant 'gaze' of health care experts, with a fundamental premise of Western medicine being the physical examination of the body of those who present to medical practitioners and an increasing amount of laboratory testing of bodily fluids. To Foucault (1973: xi), doctors and medical knowledge create their own 'rational discourse' around the body, its form and diseases. Thus:

> the presence of disease in the body ... are challenged as to their objectivity by the reductive discourse of the doctor, as well as established as multiple objects meeting his positive gaze.

The developments of modern medicine and medical perception made visible that which was invisible. That which was visible to the medical gaze and medical perception, enabled doctors '*to see* and *to say*' (ibid.: xii). Doctors now translate the inner workings of the body and interpret disease, and the body's reaction to it, into a language which they understand and were educated in. Such language, or discourse, the way of interpreting, understanding and, ultimately, treating is shared amongst those who understand it, but not with those whose experiences and understandings are different. Thus, there may be numerous discourses or discursive practises in the processes surrounding cosmetic surgery.

In cosmetic surgery the patient's body presents itself to the gaze of medicine and enables itself to be defined and analysed from the perspective of the medical practitioner. Thus, for those who seek the intervention of cosmetic surgery, the practitioner who carries out the procedures can inscribe his own notion of beauty and acceptability onto the body. In this process healthy bodies come under the medical gaze and are defined as faulty. Normal bodily characteristics can be reframed as abnormal or 'deformed', and practitioners are able to utilise the current normative standard of female beauty to reform a patient's body to conform more closely to this standard.

Dominant messages communicated through the discourse of print and electronic media mould and reinforce ideas about the body and femininity. Thus, breast size in present ideals of beauty are relatively large compared to the rest of the body and nose style and size is small and upturned. A myriad of women's magazines regularly feature articles about diet, weight loss and cosmetic surgery and promote products that defy the ageing process, aid weight loss or help keep the body trim and taut. What women understand from these messages, the way they interpret them, is individual to them as is the way they experience their bodies, their embodiment. The messages are filtered through experiences and understandings and interpreted individually in relation to individual bodies. The ways in which doctors interpret these messages are also influenced by a number of factors, not least of which is the way they perceive what women want from cosmetic surgery. This perception can be at odds with what women actually want but is underpinned by ideas of what women should look like as well as by a medical authority that defines what the 'illness' is and recommends the 'cure'.

The medical profession is in a privileged position. It has scientific knowledge and authority which puts it in a powerful situation in relation to those it treats. In therapeutic medicine, this engenders a trust in a doctor's diagnosis and treatment plan. This trust is no less real in non-therapeutic medicine, such as cosmetic surgery. However, the goal of treatment is not 'cure' but change or enhancement. The patient is there because they want to look different and the doctor is there to make them look different. While the underlying rules governing the engagement of doctor and patient are the same as in any other medical encounter, a shared understanding of what can be achieved through cosmetic surgery is influenced by 'the paternal authority of the surgeon' (Fraser, 2003b: 150) and the way that surgeon reflects female beauty.

As discussed in Chapter 3, vast numbers of women are dissatisfied with their bodies. Translating this dissatisfaction into the desire for surgical intervention, however, is not as common as taking other actions such as dieting. It is not yet clear what makes some women seek cosmetic surgery and other women, in the same situation, not. However, for those women who decide to seek out cosmetic surgery, the language used in the advertising media makes such body control and modification procedures seem straightforward and accessible. It is a discourse of beauty rather than surgery. The invasive nature of procedures and the risk attached to them is usually glossed over. Such an approach is also sometimes adopted by medical practitioners who advertise their services. For instance, an advertisement that appeared regularly in a Melbourne Sunday newspaper magazine promised liposuction that is:

> Safe, effective, same day Walk-in Walk-out procedure. All done under local anaesthetic. **Excellent Results**. Get rid of that 'hard to lose' fat in just a few hours.

Yellow pages advertisements for cosmetic and plastic surgeons invite potential patients to 'Enhance your natural beauty' and promise

'Outstanding results from affordable Cosmetic Surgery'. Many of the pictorial entries have colour photographs of scantily clad, slender and beautiful young women.

These types of advertisements also reinforce the disciplinary power of 'the gaze' both on individuals by themselves and by others. They suggest that after cosmetic surgery any woman can look like that and 'be' like that. They also suggest that the beauty is already there, it just needs to be unearthed by the doctor, quickly, cheaply and safely. Individual women can make a choice to take control and have the surgery. Such a discourse masks the reality of cosmetic surgery as a potentially risky endeavour. It also:

> constructs a preferred feminine subject that accepts direction from doctors and takes her cues from culture (for example, in the recommendation that celebrities be used as examples of what the participant aspires to look like), but sees herself as ultimately in control and independent, and accepts responsibility for the outcome of her decisions accordingly.
>
> (Fraser, 2003a: 40)

Being 'in control and independent' suggests that it is the woman who has power in this situation. However, if we adopt Foucault's ideas about power, the munificence of medical experts in providing such a service as cosmetic surgery masks the technologies of power they wield. Once the body enters the clinic it becomes an entity to be diagnosed and treated according to the advice of the 'expert' practitioner. In this way the 'docile body' (Foucault, 1979) is produced. Bordo (1993: 166), for instance, argues that:

> [v]iewed historically, the discipline and normalization of the female body ... has to be acknowledged as an amazingly durable and flexible strategy of social control.

Bordo sees that women's bodies, their shape, size and attractiveness have been controlled and affected by the panopticon of the social gaze as well as by the inner gaze and self-surveillance that women exercise on themselves. As an example of this, Bordo discusses the debate surrounding the danger of silicone breasts implants. She emphasises that:

> the most depressing aspect of the disclosure was the cultural spectacle: the large number of women who are having implants purely to enlarge or reshape their breasts and who consider any health risk worth the

resulting boon to their self-esteem and 'market value'. These women take the risk, not because they have been passively taken in by media norms of the beautiful breast (almost always silicone-enhanced), but because they have correctly discerned that these norms shape the perceptions and desires of potential lovers and employers.

(ibid.: 20)

From this perspective, women make decisions to change their body, not only because of the power of the external gaze, but because of self-surveillance and because of an assessment they make about how to increase their own power both in the market place and in personal relationships. They can be said to be exercising agency to fit into society and compete within it according to the standards and norms it (and they) have set. However, by so doing, by adhering to the rules, they are in turn reinforcing these rules, norms and standards and ensuring their maintenance and continuation.

What is in question is how much power they can yield over the final form their body will take. How much agency can they exercise in circumstances where medical practitioners interpret what they want through the lens of the power and authority vested in medicine and through their own ideas concerning what is acceptable and appropriate bodily modification? This question has been largely avoided by feminist scholars in their discussion of cosmetic surgery. They have concentrated on the motivations of women for undergoing cosmetic surgery, rather than on the role of the doctor in the process.

Feminist analysis, the female body and cosmetic surgery

Contemporary feminist debate has been engaged with women's bodies from its inception. Modern-day feminism launched itself during the 1960s through protesting at the site of the Miss America pageant in Atlantic City. Feminists were protesting against the exploitation of women's bodies through such pageants and argued that women were denigrated and objectified in this process. They also argued that beauty routines engaged in by millions of women were a product of a patriarchal society that expects women to look a certain way to be accepted. Feminists argued that the way women looked, or were expected to look, was culturally and socially constructed and was symbolic of the differential power relations that existed in patriarchal societies. As feminism found its voice, the body became a political issue with women now having control over their own fertility through the contraceptive pill and, in some Western countries, access to legal abortion. Feminist scholars have studied the many facets of women's bodily experience, including cosmetic surgery. Indeed theoretical analysis of the practice of cosmetic surgery was first engaged in by feminist writers.

Feminist scholars such as Gillespie (1996), Bartky (1990), Bordo (1993), Morgan (1991) and Wolf (1990) suggest that accessing cosmetic surgery is presented as a means through which women can increase their social power in a society where women are judged by their appearance more than men are. They also argue that women are 'targets of marketing ideology' (Gillespie, 1996) that render their shape and appearance as central to their self-identity. Youth and beauty are linked with social power and '[w]omen are constantly bombarded with images that the majority can never match up to' (ibid.: 72). While changing the way their body looks may be beneficial for individual women, Gillespie argues that the achievement of such individual satisfaction may disadvantage women in general and 'contribute to social, cultural and discursive norms that contribute to their own oppression and that of others' (ibid.: 82). At the extreme, some of these arguments are criticised as portraying women as 'dupes' of the advertising media and ignoring the very real benefits women report that they have gained from cosmetic surgery.

Wolf compares cosmetic surgery with foot-binding and genital mutilation, as practices women submit to as a result of social pressures which are, at the same time, marketed through the rhetoric of choice. She identifies the institutionalised power that lever women into making inappropriate choices about their bodies and questions whether doctors should ethically be allowed to provide what she identifies as unnecessary cosmetic operations. Here, she does not blame women for choosing cosmetic surgery but doctors for providing it. Doctors, Wolf argues, both create and service a demand for cosmetic surgery to serve their own economic interests.

Bordo (1993) contends that individual women may benefit from cosmetic surgery but, for women in general, the practice poses significant difficulties. In looking at the meaning of advertisements, she speculates that women know that the images they see are computer enhanced and otherwise perfected, but they do not internalise this at an intellectual level. These visions of the ideal limit choices about the body women can achieve.

Bordo applauds the 'old feminist discourse' (ibid.: 31) for providing a systematic critique of women's subjugation through beauty. However, she criticises contemporary feminism for being 'strikingly muted' (ibid.) on the issue of the objectification of women's bodies. What is missing in feminist analysis of cosmetic surgery is the acceptance of women's experiences of their bodies and their subjectivity. Morgan (1991) utilises metaphors of oppression consistently in her critique of women's utilisation of cosmetic surgery. She argues that resistance and refusal are a solution to the 'colonisation' of women's bodies by cosmetic surgery. However, she ignores the other influences that shape and create idealised images of women's bodies and fails to acknowledge what women themselves say about their reasons for accessing cosmetic surgery. While women in this research have reported that they 'do it for themselves', one cannot deny the influences that come to play on them. However, women's discourse surrounding cosmetic surgery has an element of resistance and refusal to accept the 'hand that they have been dealt', be it an inherited feature or the consequences of child bearing.

Lacking in Bordo's analysis is an application of her exploration of agency and women with eating disorders to women who have cosmetic surgery. She sees women who undergo cometic surgery as voluntarily offering themselves to subordination and this presents a significant tension within her analysis. There is a lack of engagement with the reality of the ambiguities and contradictions inherent in women's decisions to have a cosmetic procedure and in their seeking a 'normal' body that they can inhabit in comfort, at the particular time and in the particular social and cultural milieu that they live in.

Morgan (1991) argues that women's bodies are colonised through cosmetic surgery. She cites Foucault, and the notion of docile bodies, in her analysis of women's engagement with cosmetic surgery, and points out the disciplinary power of such an engagement over women's bodies. Morgan suggests that all women should either refuse cosmetic surgery or that beauty should be subverted by focusing the skills of cosmetic surgeons to create 'ugly' bodies. While both these ideas may be theoretically possible, they do not conform to the reality of current Western notions of beauty and, for some women, viability.

Feminist theorising has, over a period, moved from seeing beauty as oppression, a way of keeping women subordinate, to the postmodernist notion of cultural discourse. These discourses centre around the mind–body dualism, as well as the need for complete control over the body, promulgated in contemporary industrialised societies (Bordo, 1993). Davis (1993: 24) argues that this change in focus:

> enables a sensitivity to the multiplicity of meanings surrounding the female body as well as the insidious workings of power in and through cultural discourses on beauty and femininity.

However, whether the argument focuses on the oppressive role of beauty or on beauty as cultural discourse, Davis contends that the argument still revolves around the role of beauty practises in controlling and disciplining women. The dilemma, she says, is that portraying women as victims of beauty practises rather than active decision makers about their own bodies denies them power and agency.

Research conducted in the Netherlands by Davis (1995: 12) led her to conclude that:

> Cosmetic surgery is, first and foremost, about identity: about wanting to be ordinary rather than beautiful.

She goes on to say:

> Cosmetic surgery is not about beauty, but about identity. For a woman who feels trapped in a body which does not fit her sense of who she is, cosmetic surgery becomes a way to renegotiate identity through her body. Cosmetic surgery is about exercising power under conditions which are not of one's own making. In a context of limited possibilities for action, cosmetic surgery can be a way for an individual woman to give shape to her life by reshaping her body.
>
> (ibid.: 163)

In contrast to other feminist writers, Davis is arguing that women are turning to cosmetic surgery to recreate their identities rather than to create beauty. She argues that women are exercising agency in making the decision to have cosmetic surgery and that this choice should not be denied them. The act of undergoing cosmetic surgery can constitute an act of empowerment rather that subversion. While this analysis has merit, it does ignore undeniable influences that come to bear on women and how they see their bodies. Indeed, Davis glosses over those cultural and structural factors that underpin women's dissatisfaction with their bodies, preferring to add more weight to her argument that the agency of individual women needs to be acknowledged. Post-structuralist theorists such as Butler (1993) have argued in a similar vein, challenging the notions of power and oppression as too simplistic.

Morgan (1991) and Balsamo (1996) have both theorised that cosmetic surgery could be utilised for feminist purposes. They argue that the body is a cultural entity rather than a fixed natural construct. It can be refashioned and transformed in such a way as to challenge current norms of beauty. Their position echoes Haraway's (1991) conception of the cyborg, where the biological body can be transcended through technology. Thus both Morgan and Balsamo see that the body could potentially be refashioned to challenge current notions of beauty promulgated by cosmetic surgery. In the end, however, Morgan's thesis in exploring whether cosmetic surgery liberates or exploits, focuses on what she sees as 'paradoxes of choice'. Thus, women want to conform to current standards of beauty and, although the act of cosmetic alteration is a choice a woman makes, this choice is governed by that standard. Ultimately, according to Morgan, choice will become the imperative as the normal changing, ageing body comes to be seen as ugly and uncared for.

Surprisingly none of the feminist theorists mentioned here have addressed the direct role of the surgeon in the reshaping of women's bodies. Gilman (1999: 334) in his work on the history of cosmetic surgery outlines the pivotal role played by the surgeon throughout the history of cosmetic surgery in making the body socially acceptable. He says that

[w]hen we turn to the physician, we demonstrate our autonomy and abdicate it simultaneously.

Haiken (2000: 84) develops this argument through outlining the history of cosmetic surgical intervention. She argues that

facial surgery is grounded in complex questions of gender, race, culture, and personal identity.

and that practitioners who carried out such surgery were very aware of this. She links her argument to the reconstruction of the faces of First World War soldiers so that they could 'function in a society in which the male role ... was well defined' (ibid.). She also discussed the Americanisation of the face sought by immigrants to that country in the 1920s. At that time surgeons were preoccupied with 'the idea of an individual identity that was also an American identity' (ibid.) and thus they created features that best reflected the 'American face'. This direct 'engineering' of the face has meant, says Haiken, that surgeons are not just 'hapless carpenters' (ibid.: 93). They directly intervene to shape and reflect the culture of the time, a culture which:

is created each day in each surgeon's office during the series of exchanges during which patients' requests fit, or not, with what a surgeon believes to be appropriate and desirable.

(Ibid.)

Here, then, is an exploration of how surgeons create images through their own understanding of prevailing cultural and social norms of beauty and the acceptable face. This is no less true for other parts of the female body, in particular the breasts. If women are trying to recreate their identity through cosmetic surgery, they do so through offering themselves to surgeons who reflect and transpose their own understandings of culture through their scalpels.

It may be useful here to return briefly to the notion of agency discussed by Davis and others. If we define agency in this context as *control*, then the question remains as to how much control women retain when they undergo cosmetic surgery. How much influence does the doctor have on the outcome of that surgery? Can women's agency be protected once she gives her body over to the surgeon's scalpel?

In a situation where doctors have superior knowledge and authority about the body a woman's ability to retain control over her body has limitations. Medical sovereignty is a powerful force and one that can easily influence women's decisions about their bodies and how those bodies should look.

Conclusion

This chapter has set out some theoretical approaches to the body and its place in the social world which can inform and situate the practice of cosmetic surgery. While women may be under cultural, media and social influences to conform to a particular vision of 'woman', they also make informed choices about their bodies. In feminist literature, Davis (1995) argues against the contentions by Morgan (1991) that women are merely conforming to Western notions of beauty when they access cosmetic surgery: that they are being exploited by those who wield power over them (husbands, boyfriends, cosmetic surgeons) and that they are supporting their own oppression and that of other women by allowing themselves to be treated in this way. Davis criticises Morgan for not interviewing women who have had cosmetic surgery and seeking first-hand reports of why they did it. She points to the 'discomfort and unease' (ibid.: 181) that she feels characterises feminist engagement with cosmetic surgery and the fact that their theoretical approach denies women the opportunity to act as free agents and to 'take their lives in their hands'.

The arguments of Morgan, Bordo, Balsamo, Davis and other feminists, have much to offer in the attempt at a theoretical underpinning of the practice of cosmetic surgery on women. The problem is they only focus on one party in the cosmetic surgery contract – the woman. They fail to critique the role the doctor has in recreating women's bodies through surgery. How can women's agency stay intact when another reshapes her body? While some women will know exactly how they want to look after their procedure, others have a vaguer notion in that they will convey a need to 'look fresher' or 'not have such a prominent nose'. As will be explored later in this book, most women do not want their new look to make them stand out: the majority just want to blend in. This is interesting in itself, as women construct their idea of what they need to look like in order to blend in to their environment. In order to fit in and make the most of their physical selves in society, the body becomes a 'reflexive project' (Giddens, 1991).

The role of doctors in the medicalisation of women's bodies has been recognised and analysed in relation to reproductive functions in particular. In this lottery, women may get what they want, or close to what they want, or they may not. If women *do* exercise agency in making the decision to have a cosmetic procedure, the limits of that agency are very much defined by the doctor who carries it out. Once the procedure begins, woman's self-determination or agency is compromised and power over the outcome lies in the hands of the doctor.

To return to Foucault, this power is not a centralised force or something that the cosmetic surgeon purposefully wields, but it evolves from dominant historical, social and ideological ideals that inform the way doctors act in constructing and reconstructing women's bodies. At various times, and across various cultures, various types of social or moral problems have been medicalised. Normal biological functions, such as fertility, childbirth, menopause and menstruation have been brought into the remit of medical practice. In this way, the medical profession has been able to intervene in most aspects of biological functioning and govern how we experience, and deal with, these functions (Zola, 1972).

Historically, doctors who carry out cosmetic surgery have intervened at the site of the 'damaged' body to make it more acceptable to the society it exists within. Current cosmetic surgery practises performed on women continue to intervene to make the body more 'acceptable'. Doctors here become technicians of women's dreams. However, repairing a syphilitic nose or a war-injured face is aesthetically less problematic than enhancing a physical feature. In reconstructive surgery, the patient is grateful for any assistance towards normality. The patient who seeks cosmetic surgery for improvement or enhancement, however, has their own image of the way they want to recreate their body. This idea of what they want may be socially constructed and related to their own lived experience. The doctor who carries out the procedure not only works within certain technical boundaries, but also carries his notion of aesthetic beauty and how the body should look. The doctor, then, acts at the interface of culturally and socially constructed notions of beauty and the expectations of the women that they operate on.

5
Into Battle for the Cosmetic Surgery Market

As the popularity of cosmetic surgery has increased and advertising cosmetic surgery has become widespread, tensions have developed within the industry. These tensions centre around which medical practitioners should be allowed to carry out cosmetic surgical and medical procedures and which should not. These debates emanate from within the ranks of those who practice cosmetic surgery, rather than from any regulatory bodies or from the public. As noted in Chapter 2, similar tensions were experienced earlier in the twentieth century, however they have now become more acute and are particularly active in Australia, Canada, the US and the UK, indeed they positively flourish in any country where cosmetic surgery is a profitable business. The result is a 'turf war' within the industry for a share of this lucrative market. This chapter will consider these debates, and the 'turf war' that is part of cosmetic surgery, against the background of cosmetic surgery as commerce and as a practice that sits somewhat uncomfortably within medicine.

The 'Turf War' in cosmetic surgery:
Who does and who should operate?

Surgery carried out for purely cosmetic or aesthetic reasons does not attract public funding in countries that have state-funded public health systems unless the procedure can be shown to be necessary for the physical health of the patient. In some instances, breast reduction attracts a subsidy from the public purse if it can be shown that the size of the breasts is causing physical harm or significant discomfort to the patient. Most private health insurance companies do not provide cover for cosmetic surgery. Almost all expenses must, therefore, be met by the patient and, as such, cosmetic surgery is very much a service industry that can be accessed by anyone with the money to pay for it.

In Australia, like many other countries, most doctors who carry out cosmetic surgery need a medical degree but no other training is required. Currently, general practitioners, dermatologists, cosmetic surgeons and plastic surgeons carry out cosmetic procedures. For instance, in a report prepared for an inquiry into cosmetic surgery in New South Wales, Elix and Lambert (1999) in a survey of cosmetic surgery recipients found that 38.2 per cent of their respondents were operated on by plastic surgeons, 28.9 per cent by cosmetic surgeons, 9.6 per cent by dermatologists or ENT specialists, 5.4 per cent by general practitioners and 5 per cent by eye surgeons. The remainder were treated by beauticians and nurses.

The *Cosmetic Surgery Report* (Health Complaints Commission, 1999 85:5) states that

> Cosmetic Surgery is very competitive with doctors competing to establish themselves as the leading provider. Twelve specialist colleges and professional associations made submissions to the Committee on behalf of an estimated 500 doctors performing cosmetic procedures in Australia.

These specialist colleges are made up of long-established colleges, such as those representing plastic surgeons or dermatologists, and colleges created more recently to represent the growing number of cosmetic surgeons. Many of the submissions to the Inquiry reflected the debate about who should rightly be able to carry out cosmetic procedures. While doctors continue to argue among themselves about their right to practise cosmetic surgery, little attention has been paid to regulating the industry as a whole or to systematically gathering data about the number and types of cosmetic procedures being carried out. This is despite the fact that, against a trend of a fall in the frequency of medical indemnity claims against doctors in Australia since 2001, there has been an upward trend in claims against plastic surgeons and cosmetic practitioners.

Since the mid-1990s the industry is reported to have doubled in Australia and there is every indication that it is continuing to grow. The US recorded 11 million cosmetic surgery procedures in 2006, an increase of 7 per cent from the previous year (American Society of Plastic Surgeons, 2007) and we can only assume that such growth is mirrored elsewhere. It is little wonder then, that there has been a spirited debate within the medical profession about who should carry out cosmetic procedures. Plastic surgeons argue vehemently that for surgical procedures a practitioner should be a qualified plastic surgeon who has undergone training and specialised in reconstructive surgery. Others argue that for procedures such as laser treatment or liposuction, expertise in using the tools required for the procedures is more important. Sullivan (2001:147) details the rivalry that has characterised the cosmetic surgery industry in the US and argues that although 'detente' has been reached a 'formal treaty' (2001:103) is yet to be agreed. Practitioners from different medical disciplines sit uncomfortably together in an industry that is rapidly growing with new techniques and tools being developed to grow the market further.

Currently there are no specific requirements or licensing arrangements for medical practitioners who wish to practise cosmetic medicine or surgery in Australia. Of course, all medical practitioners are subject to the regulatory framework for registration and practice that applies in the state or territory in which they work. They are also subject to the common law that, in general, imposes a duty of care on medical practitioners governing their dealings with patients. The plastic surgeons interviewed for this research felt that cosmetic surgical procedures should be carried out by qualified plastic surgeons or, at the very least, by experienced surgeons. This view is supported by Marshall (1998:27) who says:

> In choosing a facial cosmetic surgeon, the patient must ultimately accept responsibility for the decision. There will always be individual exceptions, but as a general rule you need to ensure that the doctor has a recognised specialist surgical qualification, with training in cosmetic facial plastic surgery. This will usually be a specialist plastic surgeon, but may be a specialist in ocular plastic surgery, or in head and neck surgery.

Such responses were common from plastic surgeons and are reflected in the literature that they produce for patients. For instance, some plastic surgeons argue their point in publicity material. One surgeon[1] says:

> It was recently published in a major newspaper to inform the general public the difference between a Plastic and a Cosmetic Surgeon. As quoted 'any Medical Practitioner can claim to be a Cosmetic Surgeon, while a Plastic Surgeon must undergo at least eight years of **extra**, and **extensive**, training'. It is most important that you as a patient have a complete understanding of this.

Doctors in this study reflected these arguments. For instance, Dr Green, a plastic surgeon said:

> Look, cosmetic surgery now is very much like selling a used car. It is being provided by people with questionable talent, questionable training and questionable morals and ethics. And those people running huge before and after photographs with huge testimonials, advertorials etc., and once they get a client with a hook in its mouth, they will negotiate a fee ... it feels like selling a second-hand car. I would like to think I was different.

On the other hand, cosmetic surgeons and dermatologists also articulated their views about who should be allowed to carry out cosmetic procedures. Dr Apple, a dermatologist noted that

> there is a person out there practising, who gave up medicine for eight years and he had his own computer company. They went broke, and the next minute he is a cosmetic surgeon. Now I don't know what he did to become a cosmetic surgeon.

Although dermatologists do not undergo extensive surgical training, this respondent had a substantial cosmetic surgery practice and carried out surgical as well as medical procedures.

While plastic surgeons and dermatologists criticised cosmetic surgeons, cosmetic surgeons were also critical of plastic surgeons for the views they articulate about who should be allowed to practise cosmetic surgery:

> my training is extensive, but I'm not a Fellow of the Royal College of Surgeons. I have more experience under my belt than many of them. I would never do some of the things that they do, I know my limitations. Plastic surgeons get no training in cosmetic surgery during their fellowship but they do have this attitude that they, and only they, should do cosmetic surgery. It's a problem all over the world.
>
> (Dr Mile)

Marshall's (1998) view that the patient should shoulder the responsibility for the surgeon they choose ignores the fact that these patients often have little independent information about the surgery they are seeking, and even less understanding of the skills they should look for in a surgeon. It is very difficult for any patient to evaluate who is best to carry out a particular procedure when little independent information is available and when there seems to be an active tension within the field of cosmetic surgery about who should carry it out.

This debate has sometimes spilt over directly to patients. Many doctors do not seem to realise that the difference between a cosmetic surgeon and a plastic surgeon is not clear to patients. However, they hold the patient, in some way, responsible if they choose a cosmetic surgeon to carry out a procedure and that procedure goes wrong. For instance, Holly reported a less than sympathetic reaction from a plastic surgeon:

> I eventually got on to a couple of plastic surgeons. I don't really know when I worked out the difference between plastic and cosmetic surgeons, maybe when I went to see these other people and they explained to me. I found another one close to work. He was next to useless: all he did was give me a big lecture about what the difference is between a plastic surgeon and a cosmetic surgeon and wasn't I stupid to go to a cosmetic surgeon, I deserved all I got. He was very rude and I ended up bursting into tears and storming out of the place, and then he had the cheek to send me a $90 bill.

Lin said:

> When I look back I wonder why I put myself through that! Now there was definite asymmetry between one eye and the other. I couldn't stand myself. I went to see every plastic surgeon and dermatologist possible, to try to get it fixed up. I felt that I was trapped because no one would do anything about it. There was this real dislike from one group to another. Dermatologists were putting plastic surgeons down and vice versa.

There is currently no independent evidence to show whether plastic or cosmetic surgeons or dermatologists have poorer results on the whole for cosmetic surgery or any other cosmetic medical intervention. As in any profession, there are individuals that have attracted more complaints than their colleagues, but without adequate data no generalisations can be made about any particular group. Nevertheless while the debate about who should carry out cosmetic surgery continues, practitioners seem not to have engaged with some important questions concerning the industry in general. These questions include claims made by practitioners through publicity material about what cosmetic surgery can do, understanding what patients want and whether practitioners can fulfil patient expectations, and achieving a better standard of care for all cosmetic surgery patients, regardless of what qualifications their doctor has. Dr Pears put it this way:

> it is true that anyone can call themselves a 'cosmetic surgeon' and some of them are appalling. Some of them however, you have to admit, are very good at what they do. I can think of a couple of cosmetic surgeons in Melbourne who you would be quite happy to refer people to, or go to yourself. And there are some qualified people who you would never refer anyone to, but there is [*sic*] also qualified people whom you would be quite happy to refer.

Cosmetic surgery as commerce

The increase in the range of procedures offered by practitioners has brought the practice of cosmetic surgery closer to a commercial entrepreneurial endeavour. Not only do doctors operate on patients, but they inject them with chemicals, suck out fat and laser their bodies and faces. These procedures do not require advanced surgical skill and are becoming increasingly popular as ways of looking younger or thinner without radical surgery. Statistics reported by the American Society for Aesthetic Plastic Surgery (2008:449) show that the number of nonsurgical procedures, including Botox, laser resurfacing and Retin A treatments, increased by 754 per cent between 1997 and 2007. Surgical procedures increased by 114 per cent over the same period. Some of the non-surgical procedures, such as laser resurfacing, require an investment in expensive machinery. In order to recoup this investment, practitioners need to attract patients. Advertisements appear in women's magazines, the newspapers and the yellow pages and offer women, in particular, a 'quick fix' for a not so 'perfect' body. It is at this point that the distinction between medicine and commerce becomes blurred. In advertising in this way, doctors target the vulnerability of many women who see themselves as imperfect compared to the images of perfect women they see through the media. Here, the doctor is complicit in supporting the view that women's bodies need cosmetic intervention to make them acceptable, and adds legitimacy to the commodification of the female body. Miller et al. (2000: 355) see advertising by cosmetic surgeons as 'ethically problematic':

> Advertising cosmetic surgery puts physicians in the position of selling invasive procedures for which there is no medical need.

Demand-stimulating advertising is especially problematic in medicine, since the willingness of physicians to provide treatments may operate as a legitimation in the eyes of patients.

Of course, it can be argued that doctors are merely providing a choice for patients wishing to access their services. In a society where we are free to make choices about our lives, this is but another choice that we can make. However, our choices are limited by our ability to pay for the procedures and the influence prevailing social and cultural standards of beauty have on us. Those who cannot pay are denied the choice. Those who are influenced by advertising and magazine articles that discuss cosmetic procedures may be convinced that such procedures are less invasive and risky than they actually are. Such publicity also plays on many women's insecurities about their bodies or makes them feel that, in order to be healthy and beautiful, they need to change the way they look.

The more widespread availability of cosmetic surgery procedures has had an effect on the economics of the industry. Practitioners are beginning to recognise this (Krieger, 2002: 291) and address it through various strategies. One already mentioned is the differentiation of plastic and cosmetic surgeons and the pressure from plastic surgeons to regulate the market through allowing only those qualified in plastic surgery to carry out procedures. However, even if more plastic surgeons opt to carry out cosmetic procedures, the pressures on price will still exist. At the moment the 'crowded cosmetic surgery market' (ibid.:614) means that other business strategies would, according to Krieger, need to be adopted. He suggests 'three protocols for success: discounting, differentiation, and focus' (ibid.). What this means is that, like any other market with an increasing number of providers, the options are to cut the price or differentiate yourself from the competition through offering a 'boutique' service. This nexus between business and medicine is more commercially focused than in any other area of medicine and provides a unique tension. This tension has seen doctors advertise using marketing strategies that are particularly aimed at women and that are designed to make cosmetic surgery look accessible, affordable and straightforward.

The marketing of cosmetic surgery

The marketing of medical services through advertising is only a fairly recent development in Australia. Until 1994 medical practitioners were not allowed to advertise their services. Since they have been able to advertise, most practitioners list their contact details in the Yellow Pages. Those who offer cosmetic surgery have been at the forefront of more aggressive advertising in magazines and newspapers, showing before and after images of patients and inviting patients to 'enhance' their look and promising a 'safe, effective, same day Walk-in Walk-out procedure'. Other advertisements invite patients to 'discuss the way you look with someone who really knows'.

There has been some disquiet about the type of advertising engaged in by cosmetic practitioners in particular, 'before and after' photographs used to promote certain clinics and practitioners. In Australia, the New South Wales government, for instance, has legislated (2008) to regulate the use of 'before and after' photographs in advertising for medical and surgical procedures. In 2008, the British Association of Aesthetic Plastic Surgeons (BAAPS) launched an advertising campaign to counteract what it claimed were misleading marketing by some clinics in the UK.

Doctors in this research were asked about what advertising they undertook. Fourteen respondents said that they advertised and eight of these said they had Yellow Pages listings. Those that advertised more broadly mentioned newspapers, magazines, television, radio and newsletters. Others said they relied on word-of-mouth or referrals from other doctors. One respondent noted that advertising had become more important since de-regulation in 1994:

> when they de-regulated advertisement, it has gotten to a stage where you have to now turn to advertising to let people know what you do.
>
> (Dr Crisp)

Another mentioned the rise in demand for cosmetic procedures and the need to advertise to access this market.

> I think cosmetics have moved in Melbourne in the last few years and it is harder unless you advertised.
>
> (Dr Almond)

The implication here is that doctors need to advertise to access their share of a growing market. This is unlike any other branch of medical practice. While there may be a growing market in patients with chronic illnesses such as diabetes, specialists in these disciplines do not seem to feel the need to advertise their services. Indeed, the shortage of practitioners in most disciplines means that that there is more patient demand than doctors to service it. Advertising cosmetic procedures and how patients responds to this will be discussed more fully in Chapter 8.

A most unusual medical practice?

The doctors interviewed recognised that other medical colleagues often saw that the cosmetic work they did was not 'real medicine', non-therapeutic and aimed at profit. As we will see in later chapters they tend to justify their work by making it seem psychologically therapeutic and, therefore beneficial for patients. Miller et al. (2000) question the 'internal morality of cosmetic surgery:

> Although increasingly popular, cosmetic surgery is a most unusual medical practice. Invasive surgical operations performed on healthy bodies for the sake of improving appearance lie far outside the domain of medicine, a profession dedicated to saving lives, healing, and promoting health ... Is cosmetic surgery a medical privilege or an abuse of medical knowledge and skill?

Cosmetic surgery does sit uncomfortably within medicine. Those who have fewer surgical qualifications can be labelled as 'chancers' and profiteers by those who are more qualified. However, those that have undergone many years of training in plastic surgery, and associated areas, cannot escape peer disapproval and criticism about the work they choose to do:

> I think that in the mainstream of medical practice there is still the feeling that cosmetic procedures aren't really part of mainstream medicine. Because they are discretionary, they may be regarded as a little, shall we say, frivolous. Mainstream medicine deals with people who are sick or who are potentially sick and maintaining good health and that doesn't extend to alteration of appearance.
>
> (Dr Pound)

Comments were made about the conservatism of the medical profession and how this influenced their attitude to cosmetic surgery. This was seen by some as positive and others as negative. Some doctors aligned themselves with the 'conservative' pre-disposition of many doctors:

> we are traditional in medicine. There are those that will look upon this as not real medicine and they deride it.
>
> (Dr Yard)

there is no doubt that there is a very conservative element in the medical profession and that's good. This is surgery, it's real surgery and real surgery has complications and it needs to be thought of in that frame. It's appropriate that there be a cautious change in the medical profession.

(Dr Lemon)

Others were frustrated by such conservatism, seeing it as a failure to keep up with a global trend towards an acceptance of cosmetic surgery:

I think we are still in the dark ages in Australia, I do a lot of lecturing overseas, I'm in America a lot and I think that Australia still lags behind ... The medical community frowns on cosmetic surgery, they think people are stupid, they very much frown upon it.

(Dr Metre)

Negative attitudes were put down by some to jealousy both from outside the cosmetic surgery industry and within it:

I think that the medical profession are often a bit envious of plastic/cosmetic surgeons. They feel that they are making a lot of money and that it is rather frivolous and that it is not as important as what they do.

(Dr White)

Now within the medical profession there is a lot of jealousy about who does cosmetic surgery, who is qualified to do it and it is an ongoing battle.

(Dr Mile)

Other doctors who discouraged patients from undergoing cosmetic surgery were seen to be acting out of ignorance and jealousy rather than from a wish to provide professional advice to their patients:

It is changing but it is still pretty negative and it is surprising. A number of people who want to come and see me are sometimes discouraged by their general practitioner, which is based on ignorance, possibly jealousy, envy – all sorts of things like that.

(Dr Step)

Others commented that some elements in the cosmetic surgery industry had taken advantage of the opportunity to make a better living and this had led to a trend to de-medicalisation and of utilising the skills of non-medically trained staff:

> There are major problems out there in this industry. The industry has become very attractive to practitioners and not only to practition-ers but to hangers on like beauty therapists. Why has it happened? Because, Medicare through its rebates squeezed a lot of GPs and so they looked for more lucrative areas. They struggled to break into the area until 1994 when advertising was deregulated. That gave them access to what they called clients and what I prefer to call patients and under the protection of the ACCC, no one took them on and they flourished.
>
> (Dr Lemon)

In the main, the medical profession in general was seen as having a con-servative attitude to cosmetic surgery. Doctors who practise cosmetic surgery were conscious that other practitioners outside their discipline were critical of them. However, they differentiated themselves from those in their industry whom they saw as giving the practice a bad name. Here again the 'turf war' resurfaced with 'jealousy' and 'hangers on' being identified as sullying the reputation of the industry. There was little attempt by doctors to reflect on whether criticism from the general medical community had any justification, they saw such criti-cism as arising from a lack of knowledge about what cosmetic surgery can offer patients and from a certain paternalism that regards cosmetic procedures as frivolous and unnecessary.

Cosmetic surgery: Private reticence and public acceptance

Both doctors and women believed that cosmetic surgery would become increasingly popular. However, when women were asked about how they thought society as a whole currently viewed cosmetic surgery many felt that the view was a negative one.

> I think most people probably still disapprove a bit. I know a couple of my friends, when I told them, they were quite shocked.
>
> (Molly)

Some women reported that they felt others judged them as vain and were disapproving:

> My close friends know I've had it, but most people would think it wasn't necessary, it was bit of a waste of money and that you are vain.
>
> (Chloe)

I don't know. Some people are disgusted. We were up front about it so people were OK but some people feel that if you don't accept yourself as you are then nobody else is going to.

(Charlotte)

However, others felt that the reaction was mixed:

Friends that I've told, it was a case of 'good on you. You look fantastic.' They admired my courage to do it. I haven't told family because they would consider it a total waste of money – that could come off the mortgage. Outside of that it's no-one else's business.

(Donna)

Everyone has a different opinion, you're either for it or your not. My parents were against it because they were worried, but they knew that I would get it done anyway. I told all my friends and work colleagues and I tell people now. It hasn't worried any guy that I've been out with.

(Fiona)

Some women were reluctant to tell friends and, in particular, family about their surgery for fear of disapproval. Most of the women had been selective about who they chose to tell about their surgery:

I had told a couple of my friends about my plan but no I would never tell my mother: she would be horrified.

(Sonia)

Others detected that societal attitudes were changing and that cosmetic surgery was being regarded in a more positive light. It was suggested that this new attitude was driven by an expectation that we all need to 'look perfect' and that, in future, men would also come under this pressure and, increasingly, undergo cometic procedures:

Over the next 5 years I think there will be more procedures carried out. They are softening to it because there is more of it. It's just becoming more acceptable, it's a good thing. We're so Americanized that that's just the way its going. There's this push that everyone has to look perfect. Really, if you look good you get a lot further in life, basically, it's a sad fact but it's true. There's more pressure on women than men to look good.

(Tanya)

I think now it is changing because men are starting to do it. Now that the pressure is also on men – now we're starting to see good looking guys with good abs and six packs and good bodies and faces. So now I'm actually glad that it's having this comeback upon men, and it's not just women any more that are under pressure – it's the men now.

(Hope)

Conclusion

There are significant and deep-seated tensions in the cosmetic surgery industry. These tensions are embedded in the attitudes of doctors who carry out this surgery towards colleagues in their discipline and in the way the medical establishment in general views those doctors who choose to practise cosmetic surgery. Although significant confusion exists among women about practitioners and their qualifications, regulators have avoided any particular regulation of the industry. This is surprising since doctors who work in this area function within an industry that sits astride medicine and commerce. No other area of medicine relies so heavily on marketing for survival while, at the same time, responding to culturally constructed visions of women's bodies. It both responds to, and helps create, prevailing popular visions of women.

The active and ongoing debate within the industry about who should be able to practise cosmetic surgery is noteworthy. It draws attention away from critiquing cosmetic surgery as a practice and is not well understood by the consumers of cosmetic surgery. Although critical of the commercial behaviours of some practitioners, there is little attempt by any group of practitioners, be they plastic surgeons, dermatologists or cosmetic surgeons, to disengage from the commercial promotion of the alteration of women's bodies for cosmetic reasons. Indeed, doctors are actively engaged through the media in promoting and supporting cosmetic surgery and in putting on display their expertise. At the same time, they are critical of the media for the role they have in objectifying women's bodies and in creating visions of women that fall well outside the aesthetic reality of what most women can achieve. Through their engagement with the media, doctors not only promote surgical alteration of women's bodies but are actively engaged with the socio-cultural system that objectifies those bodies. Being themselves embedded in this context, in turn, influences their perceptions of women's bodies and makes questionable any objectivity that they purport to have in the way they sculpt the body during surgery.

Tensions about cosmetic surgery also exist in the wider society. Because of this, women in this study were often secretive about their surgery, choosing not to tell too many people about their decision. Most women felt they were regarded as 'vain', 'silly' and 'egotistical'. This contrasts with what they actually felt, that is, as needing to solve a problem with the way they looked. This could be re-conceptualised as taking control of their bodies by choosing cosmetic surgery rather than being passive victims who accede to cultural pressures. What is clear from what women say is that they felt surrounded by ambivalence about cosmetic surgery. Doctors felt the same tension, particularly from within the medical profession as it is a difficult area of medical work to justify to their colleagues.

Despite all of this there is little doubt that more women, and a more diverse group of women, than ever before are choosing to undergo a variety of cosmetic procedures. Even if patients believe societal attitudes are conservative, all forms of media continue to be a powerful influence on the development of the industry. While women may be criticised as vain by society on the one hand, on the other they are encouraged to take control of their bodies and invest in it by the media and in the advertisements by practitioners. It is this atmosphere of tension and confusion that provides the backdrop to women's engagement with cosmetic surgery.

6
The 'Why' of Cosmetic Surgery: Patient and Doctor Motivations

As we saw in Chapter 4, there has been some theoretical debate and discussion about why women decide to undergo cosmetic surgery. However, very little empirical work has been done in the social sciences which actually asks women themselves about this. An exception is the work carried out by Davis (1995) in the Netherlands which shows women as wanting to fit in and feel normal rather than stand out for having, what they perceived to be, flaws in their body. Their decisions concerning cosmetic surgery were not motivated by a desire to be outstanding or beautiful but by a desire to feel at ease with their own bodies. As cosmetic surgery becomes more normalised and easily available to a broader range of women, it is important that we understand why women opt for cosmetic surgery and what influences the decisions they make. Rather than being unusual, cosmetic surgery is becoming increasingly common and is being utilised by many women to achieve self-transformation (Heyes, 2007). If we can understand the stories that lie behind cosmetic surgery, then we can begin to see what motivates women to undergo such risky and significant changes to their bodies and what self-transformation means to them.

Absent from any theoretical debate about cosmetic surgery is an examination of why doctors choose to practise cosmetic surgery. While we may be able to surmise that doctors enjoy this type of medical work and find it profitable, we need a better understanding of their motivations if we are to appreciate the dynamic that informs their work and their relationship with the women they operate on. This chapter will delineate women's stories of why they embarked on the journey of cosmetic surgery. It will also present doctors' views on why they think women come to them for cosmetic surgery. It will outline how doctors explain the choice of cosmetic surgery as their preferred discipline and, again, in contrast, describe what women see as the doctors' motivations for working in this area.

Why women want cosmetic surgery

The decision to undergo cosmetic surgery is not one that is taken lightly by women. The women interviewed in this study had often deliberated long and hard about the choices they were about to make about their bodies. Many women had been considering surgery for a long time before they took any concrete action and contacted a surgeon. Although the surgery women wanted to undergo varied, their explanations about the feelings they had about their bodies were remarkably similar. They talked about the particular feature that bothered them almost as an intruder on their body, something that did not fit with the rest of their body and did not reflect who they felt they were. Women justified cosmetic surgery as a necessity for themselves rather than a choice. They acknowledged the tensions surrounding their decision, but saw themselves as somehow different from those they saw as having frivolous reasons for undergoing surgery. Their decision to proceed with surgery was often made alone. Most women told only their closest friends and family about this decision and many chose not to tell even close family.

The women in this study gave three broad reasons for wanting to have cosmetic surgery. These were:

- It was something that had bothered them for a long time, such as a large nose;
- It had come about due to changes in the body due to pregnancy, weight loss or gain and they wanted their old body back or;
- It was a result of ageing. When they looked in the mirror they didn't look as they felt. They wanted to look fresher either for themselves or because they felt they could compete more effectively in the labour market.

When women talked about why they wanted to undergo cosmetic surgery, they described how dissatisfied they were with a particular part of their body. They were not dissatisfied with their body as a whole, but felt that a particular feature let them down. For some women their unhappiness had begun as children or adolescents and was linked to being teased. Others described that they felt cheated because their bodies had changed as a result of pregnancy or weight loss. They had 'done the right thing' and breastfed their children and were left with shrunken, saggy breasts. Women who were concerned about the consequences of ageing wanted to be better versions of themselves rather than to look different.

When women talked about the nature of the changes to their bodies, they were clear that they did not want to be noticed. They did not want to stand out; they wanted cosmetic surgery to enable them to fit in to society. That errant part of their body that needed changing was seen as a barrier to them projecting who they really were. They wanted their bodies to meld seamlessly into its surroundings, not to stand out in it.

A long-standing problem

Women spoke of having a feature, a long-standing problem, which had 'bothered' them for a long time and caused them to feel bad about themselves. These feelings began during the teenage years or before. It was clear from what women said that their feelings about this feature were deep-seated and caused them significant anxiety over a long period of time. Over the years they had adopted various strategies to take the focus away from the problematic feature. For instance one woman talked in detail about taking various steps to change the way she looked so as to detract attention from her face which she felt was her real problem:

> I felt bad about my face since teenage years, probably around 16. I went on diets, liver cleansing diets and tried to make my hair look nice and things like that, but it was the face that was the problem.
>
> (Chloe)

Another woman had felt since childhood that her nose did not fit in with her face:

> I was a bit paranoid about my nose and I didn't like to see myself in silhouette, it seemed to be the one thing about myself that was out of proportion with the rest of my face so after giving it a lot of thought I decide to do it. Ever since I was a youngster I always thought my nose was big.
>
> (Charlotte)

Some women were self-conscious because they had been teased by their friends and peers:

> I had a bump on my nose ever since I was a kid and the other kids used to make fun of me. I remember always being self-conscious

about my looks. I also had psoriasis all over my body. I couldn't change the psoriasis but I could change the nose.

(Jodie)

Had I not been teased in high school, I might not have thought anything of it but it made me self conscious.

(Greta)

I had a rhinoplasty done because I had a hook in my nose, and I used to get teased so I got that done.

(Hope)

The emotional consequences of such teasing had created significant dissatisfaction with their body that had remained with them from their youth.

Some women had inherited this feature from a parent. This caused particular anxiety as it was linked to their heritage and identity:

There was this one thing that bugged me, the size of my nose and the shape of it. It was such a family trait so people who would meet you for the first time would say 'Wow you've really got your Dad's nose'.

(Charlotte)

I was never happy with my nose, it was too big and I was not happy with my appearance. Everyone said I looked like my father and my father's got an awful nose.

(Greta)

Both these women had noses that were a sign of their ethnic and family heritage. By having their noses changed, they were also acting in defiance of this heritage and lineage and had to deal with the reaction of their families to their decision.

Looking at a parent and knowing they had the same feature was a constant reminder to the following two women that their bodies would not change unless they took action. They had been unhappy with these features for a long time and were determined to change them when they were able to:

It's always been something that I've been unhappy with. Unfortunately it runs in the family. My mother's exactly the same. So we're very much the classic pear shape, so from the waist up we're a size eight. We seem to hold fat in that area. It's something I'm always going to be unhappy with and since I first heard about this, which was when I was quite young, I can always remember saying ' I don't care how much it costs. One day I'm going to have it done'. I was very determined.

(Freya)

Because I was always conscious of my nose, my Mum's got the same nose and I have been conscious of it all my life. I didn't say anything because I didn't want to hurt my Mum's feelings, but I had the opportunity to have it done so I did it.

(Grace)

Women who had such long-standing problems only wanted to deal with a particular feature and remove the burden that it placed on them. For some of these women the decision to have cosmetic surgery was complicated by the fact that they didn't want to hurt a family member's feelings or feel that they were denying their heritage. Resemblance to family is very much part of all our identities, our sense of who we are and of belonging. It was clear from talking to these women that changing an inherited feature was a significant step for them to take. In some ways taking such an action to reform themselves was also denying a formative part of themselves.

One woman mentioned that her rhinoplasty was interpreted as not accepting herself for what she was:

Some people feel that if you don't accept yourself as you are then nobody else is going to.

(Charlotte)

It is clear from these reports that women had considered cosmetic surgery over a number of years and had dealt with the emotional consequences of being constantly self-conscious. Added to this, some women felt they had to deal with family disapproval or hurting a family member's feelings by undergoing cosmetic surgery to change an inherited feature.

The consequences of child-bearing

There were other women who had not been concerned with their bodies or how they looked until they experienced bodily change after child-bearing and breastfeeding or weight loss. These women wanted to reclaim their 'normal' bodies. They talked about 'being cheated', as if their body had been taken from them and replaced with another:

> I just felt as if I had been cheated of my normal body by pregnancy and delivering the children.
>
> (Clare)

They were so self-conscious about their bodies that they did not want to be naked even in front of their husbands:

> And then after I had four children and breast fed them, then they also drooped so I just felt self conscious about them. Even with my husband because I had a lot of loose skin. It just didn't look very nice.
>
> (Irene)

> My breasts were really saggy after pregnancy and breastfeeding and I was very flat-chested. It was terrible, I was very conscious of it. I wouldn't take my top off in front of my husband. I felt very ugly.
>
> (Thelma)

Women described their breasts as 'sacks', 'saggy' and 'deflated balloons'. It was as if they had ended up with something that wasn't really theirs:

> After I breast fed my first baby my boobs were just flat, just sacks, there was nothing there.
>
> (Jodie)

> I was 25 and having breast fed four children and my breasts were definitely a negative part of my body image. I had always had really large breasts and all of a sudden had deflated balloons.
>
> (Stella)

And they wanted to reclaim the body and breasts that had been theirs before they had children:

> I'm not going to have any more children and I wanted my chest back.
>
> (Lorna)

Here, women talked about recovering what they had lost rather than enhancing a physical feature. They felt that they had 'done the right thing' by breastfeeding their children and should not suffer because of it. They felt cheated and unattractive and were uncomfortable in their own bodies and estranged from them. Their descriptions of being 'cheated' and ending up with bodies that they could not bear to look at are compelling. They yearned to retrieve the body that was theirs before they had children.

Ageing

The third set of reasons given by women for accessing surgery had to do with ageing. They did not feel they were the person looking back at them from the mirror.

> And that's what a lot of women do complain about. They look in the mirror and it's what they feel and how they feel, they're looking at somebody who is a lot older and angrier or tired and you don't necessarily feel like that.
>
> (Kerry)

For this woman, her sense of identity was at odds with the reflection in the mirror. It also projected an image that was at odds with the reality of how she felt. Even though she did not did not feel angry or tired, her face suggested she did.

Some women felt that they needed to undergo surgery to enable them to compete successfully in the workplace. They saw it as a necessity rather than a choice. In order to feel confident in the workplace, they needed to look better. They felt that their very survival as wage-earners was linked to the way they looked and they saw their decision to get cosmetic surgery done as a rational and affirmative response to an economic necessity:

> I wanted it done because I had to stay in the workforce for another 5 years or more and I was starting to look untidy around the face. I had too much loose skin and I was just looking tired.
>
> (Dora)

> Because I wanted to look better. I am a professional person, in my workplace I work among a lot of younger people.
>
> (Emma)

While they talked about wanting to feel more confident, they did not say that they wanted to look younger, they just wanted to feel good about themselves:

work has got a lot to do with it. My job is in the fashion industry, I think my credibility is better as a salesperson if I look, well, good. I'm in the business where people look good. And also I like things to look good. I usually felt pretty good about myself, in a way. I didn't do it to boost ego or confidence, it was just sort of a necessity really.

(Ella)

It affected my self-confidence, with my employment because I didn't feel so crash hot about my appearance anymore, it just affected my self-confidence within the work place and also socially. I thought, I don't have to feel this way, I want to do something about it.

(Ivana)

Other women felt that their ageing body did not reflect the way they felt on the inside. They also felt that they needed to keep pace with a society that values youth and fitness:

I think it's a change in society, where people are more active and are feeling younger inside. They want to have the shell resurfaced to keep up with feeling young inside. I sometimes go in to shops and I look at things and I think 'that's great' and then I'm thinking 'hang on should I be wearing that?' Inside I am still probably ten years younger than my current age.

(Donna)

There was also a negative reaction to ageing and its consequences:

I didn't want to get old and the first time I saw the surgeon he asked 'What do you see when you look in the mirror?' And I said wrinkles and I don't want wrinkles.

(Ffleur)

As more women engage in the workplace and lead longer and healthier lives, there is a tension between the way they feel about themselves and the chronological age of their body. They want their body to reflect how they feel and not how old they are.

The physical feature or perceived failing that led women to have cosmetic surgery made many of them feel self-conscious and lacking in self-esteem. In their minds, it made them stand out when they wanted to blend in to a society that values and applauds the healthy, attractive body. These feelings about their bodies were reported by women irrespective of the cause of the problem identified by them. Three main categories of feelings, that had to do with presenting themselves to the world around them, reported by women were: blending in, being self-conscious and enhancing social confidence.

Fitting in

The reasons women gave for wanting to change something about their bodies were underpinned by their need to fit into the society they lived in. Contrary to some popular ideas about cosmetic surgery recipients, these women did not come across as exhibitionists, unduly vain or as wanting extreme intervention but as wishing to be 'normal' and to blend in to their surroundings. While these needs may be culturally constructed and influenced by a society that sees the body as a commodity, they had very real implications for the way these women felt about their interactions in that society. In their need to be 'normal', women wanted any cosmetic intervention they had to look 'natural'. While they did not explain what 'normal' or 'natural' meant to them, they were keen not to look very different or stand out.

Being self-conscious and wanting to blend in

The theme of feeling self-conscious and just wanting to blend in recurred throughout the interviews with women. They felt they were the object of 'the gaze' of others when they didn't want to be. They sought surgery to blend in and to be 'normal' rather than to stand out. They wanted to be like everyone else. They did not want to always be adapting their behaviour towards others so as to minimise the attention paid to their 'flaw':

> I felt I had to physically disguise my appearance in many ways by being a bit more subdued and by not looking at people. Even though I'm fairly confident, I just felt that sometimes people would look at me twice because I looked the way I did.
>
> (Megan)

Feeling that the way she looked made her stand out impacted on Megan's self-esteem and on her ability to function well in society:

> And I also felt that people were looking at me, and I was a bit of a recluse because of the dark (under the eyes).

Women had been constantly self-conscious of the feature that they had changed through cosmetic surgery. For instance, being conscious of a facial feature that they thought others were noticing at all times:

> I was always self conscious of it, particularly of my profile, I didn't like driving up to traffic lights in case people were looking at my profile.
>
> (Grace)

> I felt self conscious about the fact that my head was turning and shaking and I felt more self conscious about this bit wobbling down the bottom.
>
> (Ella)

Women's breasts, either because they were deemed to be too large or too small, were also a source of anxiety. The following account details how one woman was self-conscious about her large breasts from an early age. She was teased and she could not get clothes to fit well. When she became a mother her breasts continued to be a source of concern:

> Well I guess it affected me more when I was younger, you know boys making remarks about me and calling me things like 'Jane Mansfield' or something stupid like that. And just getting clothes to fit and having to wear these big bras that sort of weren't very attractive ... So I was always a bit self conscious about that. And then after I had four children and breast fed them, then they also drooped so I just felt self conscious about them.
>
> (Irene)

Smaller breasts were just as problematic as larger ones and they too had to be hidden from others and impacted on self-esteem:

> I had low self-esteem. I wasn't confident about meeting anybody. If I did meet anyone I wouldn't take my top off in the bedroom. If they did stay the night I'd wear a bra to bed.
>
> (Lorna)

These women seemed to be in constant battle with the part of their body that troubled them and went to some lengths to hide it. They felt they were under the constant surveillance of those around them because they thought they were flawed. They did not complain about their whole body, only the one part that caused them the anxiety and made them ill at ease when facing the world.

Enhancing social confidence

The effect on self-confidence was mentioned both in social situations and in the workplace. Women internalised the feelings of being 'physically flawed'. They felt they needed to improve the way they felt about themselves which, in turn, would improve their social confidence:

> It wasn't really damaging my social relationships but I was carrying around a feeling I was flawed. I wondered if my confidence would improve if I did something about this perceived flaw.
>
> (Hannah)

> it just affected my self-confidence within the work place and also socially, I thought.
>
> (Ivana)

Direct comments on women by others also affirmed the fact that their 'flaws' were being noticed, that they were not 'normal', and that they stood out:

> people would make comments to me like, 'Are you tired or have I upset you? You look angry. Have I upset you?' and I wouldn't be but it was just because of my face, I looked angry.
>
> (Kerry)

Gagne and McGaughey (Gagne & McGaughey, 2002: 816) claim that 'it is through looking that we are constituted as subjects and objects'. Thus, while the women interviewed stated that they underwent surgery for themselves, the gaze of others was a significant issue for some. This gaze was not identified as a specifically male gaze but more generally as a societal gaze. These women wanted not to be 'looked at'. They wanted relatively small changes to help them fit in rather than stand out. Here again we engage with the theme of normalisation and how such a concept is constructed. Gagne and McGaughey argue that there is a need to understand 'how people are pressured to become normal' (ibid.: 819). It is not an overt pressure but, as has already been argued in this book, women are 'guided in their willing obedience' (ibid.)

by the socially and culturally constructed norms of beauty that they internalise.

Women in this study talked about wanting to be normal. They either wanted to regain a body that they felt had once been normal or gain normality through cosmetic surgery. For instance, Clare felt:

cheated of my normal body by pregnancy and delivering the children.

Hannah said that, after having rhinoplasty, she felt:

quite spectacularly normal.

If it is the need to be 'normal', to 'fit in', that leads women to the clinic then how are these motivations interpreted by the doctor?

When doctors and patients come together the outcomes women expect can be very much at odds with what doctors can provide. While it is fair to say that doctors want a good outcome for their patients, the understanding of what this 'good outcome' might be can vary significantly between doctor and patient.

It is quite difficult for doctor and patient to achieve a shared understanding of precisely what aesthetic outcome a woman wants. One of the barriers to achieving this understanding is misunderstanding why the woman is there in the first place. Added to this is the fact that women often go into the process of cosmetic surgery armed with little knowledge of what it involves, what can actually be achieved from it and, little information about the practitioner involved. Their pathways to the clinic are varied. Many of the women interviewed relied on the Yellow Pages or newspaper/magazine advertisements to find their practitioner. Some were referred by friends or their family practitioner. Many commented on the lack of independent information available about practitioners, their specialism or the procedures they offered. Thus, they went in to the process of cosmetic surgery with little information, little independent or verifiable knowledge about their doctor or what procedures they were skilled at, but with an overwhelming feeling of needing to have a physical failing corrected as secretly and quickly as possible.

While women clearly articulated why they wanted cosmetic surgery and how they felt about their bodies beforehand, they had little or no understanding whether their motivations were understood by the doctor, nor how those motives would be interpreted by the doctor.

Why women want cosmetic surgery:
The doctors' perspectives

Doctors were asked why they thought women wanted cosmetic surgery. Their responses centred around psychological factors pertaining to individual women, the need for women to feel and look 'normal' and, more generally, socio-cultural factors.

Psychological factors

A recurring theme in doctors' explanations of why women undergo cosmetic surgery is linked to psychological factors. Doctors explained that what they were doing was improving the psychological health of patients through cosmetic surgery:

> what we have to deal with, when one does aesthetic surgery, is all the psychological component of it as well and it really is why people are choosing to have it.
>
> (Dr Metre)

It should be noted here that only one doctor reported using psychiatrists or psychologists regularly to assess the mental health of their patients. The others used their own judgement to ascertain whether women had a psychological disturbance. This is problematic, since they usually have no training in psychological assessment and have to rely on intuition rather than clinical assessment. One doctor commented that

> I don't refer to psychologists very much, you'd be surprised how little. I do, but not very often, no more than twice a year. If I thought someone was psychotic or dysmorphic I would do something about that. I would try to point out what a dangerous condition it is.
>
> (Dr Apple)

Another doctor was confident that cosmetic surgery was more successful than psychological intervention could be:

> but I will show you some photos of a woman that wouldn't go out and we have changed her life by improving her appearance and there is no way psychology would have solved her problems.
>
> (Dr Mile)

This emphasis on cosmetic surgery as a psychological tool by doctors who carry out cosmetic surgery is not new (Haiken, 1997) and is seen as 'life changing' for women.

Miller et al. (2000: 359) suggest that if cosmetic surgeons believe that they are treating psychological problems, then:

> in order to provide minimally competent care, they ought to be working in tandem with mental health teams ... and offering non-surgical options to at least some of their patients.

A study by Ozgur et al. (1998:419) found that

> A person who seeks an aesthetic procedure should not be considered as a psychologically disturbed individual at face value although there may be some as a subgroup, so each case should be evaluated individually and thoroughly in the preoperative consultation.

Furthermore, the study reported other work in this area that concluded:

- Of 50 female facelift patients, there were none who were psychotic and only one had a neurotic diagnosis.
- Aesthetic surgery patients tend to have higher self-esteem than the general population and

> a patient seeking aesthetic surgery is a normal individual who is in search of reducing the inconsistency between general and specific body-part esteem.

> (Ibid.)

While saying they offered a psychological service to patients, the doctors interviewed were careful to exclude patients who they thought would be 'difficult', that is, they thought the patient was psychologically unstable and they might be the type of patient who might litigate.

Fraser (2003b: 126) in her analysis of medical and non-medical literature on cosmetic surgery, identified that

> [t]he definition of cosmetic surgery as 'psychology with a scalpel' appears frequently within the literature.

She also asserts that there is a trend within the medical literature to identify dissatisfied patients as being mentally unstable. Thus, when a surgeon believes that the outcome of surgery is successful but the patient does not, the patient is labelled as unstable. Litigation is seen as further proof of such instability.

What we can glean from women's reports of their motivations is that they are 'normal' women who want to stay 'normal'. That is, they want to fit into society rather than stand out in it. This need to fit in was recognised by some doctors:

> And it is this need to conform … almost like there is a hunger drive – a drive to be similar. And that drive is probably what keeps society fairly cohesive. If everyone went and did whatever and wanted to be a total non-conformist we probably wouldn't have much of a society.
>
> (Dr Dance)

It would be untenable to argue that no cosmetic surgery patients have a psychological disturbance. One study (Sarwer et al., 1998a) found that 7 per cent of cosmetic surgery patients studied met diagnostic criterion for body dysmorphic disorder. A review of research into the psychological characteristics of cosmetic surgery recipients found results to be contradictory (Sarwer et al., 1998b: 13). The paper notes that

> the socio-cultural influences on body image may be the most relevant to understanding the role of body image in cosmetic surgery.

Such research indicates that psychological influences are not as significant for cosmetic surgery as many doctors suggest. However, the evidence in this area is equivocal. The women in this study saw themselves as being engaged in an exercise of normalisation. They may do this for functional reasons, such as competing in the workplace, for reasons of maintaining their identity and reclaiming their body, or to change a feature that made them stand out. These women may be controlled by various social institutions who define what is 'normal' for women. But they are trying to manage a body that they feel does not fit well into society and does not reflect who they are. Foucault's (1973) argument that 'bio-power' is an influential method of control over the body has resonance here as does the argument that medical professionals, through rituals of body surveillance, construct beliefs about normality and abnormality.

Normal and abnormal

One way in which medical and lay discourses in cosmetic surgery are different is through redefining features women dislike as 'abnormalities' and 'deformities'. Women themselves did not use this terminology to describe the feature that bothered them but doctors did:

> Now other people might not even perceive the deformity that my patient thinks they have.
>
> (Dr Crisp)

> I mean basically they see whatever they are concerned about as a deformity and it is affecting them psychologically.
>
> (Dr Sand)

> Some of them, they want something that is 'abnormal' changed, or that they want something that is fairly normal enhanced or that they just want to feel better about themselves for some ... they might want to be in a wedding photo or something. Or they've got a problem on their face and they think? 'I hate that – I want to get rid of it'.
>
> (Dr Pears)

By using this language doctors are reinforcing cosmetic surgery as a psychological as well as a physical intervention. They are identifying the work they do as correcting malformation rather than improving or changing a normal characteristic.

At the same time they talk about the need for women to feel normal:

> I think that they want to feel better and they also want to feel 'normal' within their group. So I think they perceive that there is something about them that sort of separates them from their peer group or whoever is with them.
>
> (Dr Almond)

> No one says I want a beautiful nose – I just want to look normal – this is what it is about.
>
> (Dr Dance)

While the language of normalisation was common to women and doctors, the identifying of features as deformities and abnormalities were not. Doctors tended to use language that categorised cosmetic surgery as addressing dysfunction or disorder, while women talked in more straightforward terms about changing a feature that they did not like and that bothered them.

Socio-cultural influences

Doctors were aware of the socio-cultural influences on women and what it means to women to look normal in contemporary society. They recognised the role the media plays in influencing the way women see themselves and the pressure that this puts on women:

> I think that there is an innate consideration of what looks good or what looks acceptable and then there are external influences, things that people are subjected to from the media.
>
> (Dr Pound)

> It is the magazines – the 'New Ideas' of the world and the 'Cleo's', there is such an emphasis on it. The thing is written up in every magazine and every issue has something on cosmetic procedures and there is pressures on women to carry out these decisions and to make themselves 'look better'.
>
> (Dr Apple)

Magazine and film images influence not only the way women feel about their bodies, but also we can speculate that the same influences come to bear on their doctors and the way they view women's bodies. Many doctors write for, or are mentioned in, magazine articles. Such publicity advertises the services those doctors offer and adds medical weight and authority to those articles. Thus, on the one hand, doctors are willing to be part of the process of media publicity, on the other, they criticise the media for its influence on the way women feel about their bodies.

It can be argued that doctors who carry out cosmetic procedures are not only responding to market demand for the service, but are, through advertising, doing what all commercial ventures do and creating that demand. They are also replicating and further supporting the idealised form of female beauty through the procedures they carry out. It is ironic that some health professionals, when dealing with some aspects of women's body image such as anorexia and bulimia, are working to counteract the influence that social, cultural and media pressures have on women, while others, through cosmetic surgery, are carrying out procedures that maintain and support these pressures.

Doctors as gatekeepers

Doctors also act as gatekeepers for women who want to access cosmetic surgery. Doctors clearly articulated a range of reasons why they would refuse to carry out a procedure on a woman. These included the woman's motivation, her expectations and her psychological status:

> Well I mean the major one is what are the motivating factors for having it done. The second one is totally unrealistic expectations of what can be achieved and the third one is psychological instability and that is often hard to establish in one interview.
>
> (Dr Sand)

One doctor mentioned, with some frustration, that cosmetic surgery was now so widely known about that there was an expectation that anyone could access it:

> Now everyone wants cosmetic surgery. Girls on social security want their breasts done and they expect the Public Hospital system to do it.
>
> (Dr Black)

The success of the cosmetic surgery industry in promoting itself has made it more accessible to women but also seems to encourage a wider range of women to seek cosmetic surgery. It is not surprising then, that doctors situate themselves as gatekeepers of cosmetic procedures.

The right motives

Doctors stated that they needed to be satisfied about women's motives for wanting cosmetic surgery. In particular, they wanted to ensure that women were 'doing it for themselves':

> I try and elucidate what their motivation for surgery is and I think that if it is not a personal motivation, if there is some external pressure or force being applied, then I don't think that they're candidates for surgery.
>
> (Dr Pound)

> One of the things you have to try to steer clear from are the ones who have met a fellow last week and he says they need a breast augment and he is going to pay for it, that's not good.
>
> (Dr Apple)

Some women mentioned that doctors had asked questions that sought to ensure that women were 'doing it for themselves' and not for a male partner:

> Yes, basically he was trying to establish that I was doing it for myself. He asked me personal questions about my marital situation and all that, so that I was doing it for me. He just wanted to establish for himself why I was doing it and that I could deal with the aftermath of it.
>
> (Ffleur)

> When I went to have the breast reduction I knew that he was saying, 'Why are you having this done? Are you having this done because of the man in your life?' And I said no, and he said, 'Well I'm glad you said that because had you said you were doing it for that reason I would have decided not to do it for you'.
>
> (Ella)

In clarifying that women are 'doing it for themselves', doctors are supporting the repertoire of agency that permeates the process of cosmetic surgery. However, they seem careful to control this agency by acting as gatekeepers to those they deem to be seeking cosmetic surgery for the wrong reasons.

Fraser (2003a: 40) identifies a trend in medical texts to promote an individualist account of agency while encouraging women to look in magazines and at the photographs of others to find their ideal. While women should be 'doing it for themselves', they are encouraged to seek out and emulate the features of others through cosmetic surgery. While doctors say they are careful to ensure women are doing it for themselves, the discourses of medical texts:

> utilise and produce a version of femininity that combines determination and strength of will with a willingness to be directed by the surgeon.

Thus, while doctors may be keen to ensure that motivation lies within the self, they ensure that control of the process lies with them. Women can only be self-directing up to a point.

Unrealistic expectations

Another reason for not agreeing to operate on a woman was what doctors described as 'unrealistic expectations'. Doctors seem to identify what 'unrealistic expectations' were on a case-by-case basis rather than by any systematic psychologically based measures:

> But if I get the impression that they're not listening and that they do have unrealistic expectations, then I will suggest then they should go away and think about it again and come back and see me once more. And if they just don't come around to understanding what can be achieved then I won't operate on them.
>
> (Dr Pound)

> I think if the procedure won't give the result the patient desires or requires and I figure that if I feel the patient has too high expectations and is not realistic about understanding what is involved in the surgery.
>
> (Dr Dance)

> If I think the patient has unrealistic expectations. Again, if I don't think I can give them what they're after or if I don't think that they know what they're after, then ... I will never be able to please them.
>
> (Dr Crisp)

Here, doctors are very clear that the woman's articulation of what she expects from the surgery must match their own, which is deemed 'realistic'. Women are expected to accept direction from the doctor and to give control to him if they want the procedure to go ahead. Self-determination for women is limited to what the doctor will allow and women are deemed as deserving cosmetic surgery only when their ideas about outcome are in tune with those of the doctor. That is, they must agree to an outcome fashioned by the doctor and not one that is solely desired and fashioned by them.

The deserving patient

As has already been mentioned, some women felt cheated of their real body and felt they deserved to get it back. This was reinforced by some doctors who asked women about their body maintenance routines:

> He wanted to know whether or not I'd tried exercise and dieting and things like that. So he wanted to know about my life-style, so obviously I don't think he viewed it as a weight loss procedure.

I explained to him that I wasn't looking at it as a weight loss procedure either, it was more of a 'get back to a shape' procedure.

(Donna)

Here doctors reinforce the message that if a woman has made an effort to diet and exercise then she is deserving of surgery. Thus women who conform to prevailing cultural values and work to tame and control their bodies are seen as deserving of the help of the doctor.

Psychological instability

It was noted earlier in this chapter that doctors claimed that cosmetic surgery offered psychological benefits to women and, as such, was medically legitimate. Fraser (2003a: 35) notes that published medical materials display:

the repertoire that links cosmetic surgery with psychological treatment and casts participation as the conscious attempt to intervene in, and attempt to improve, one's own mental health.

However, the doctors interviewed for this research were careful to select patients who they felt were psychologically stable. While they were not keen to operate on patients who they felt were not psychologically stable, they did not seek an expert evaluation of their patients' mental state:

Patients that are clearly psychotic and have mental illness. Patients who are perhaps excessively obsessional and dissatisfied with very minor imperfections.

(Dr Almond)

Sometimes I say your personality is not suited to plastic surgery you are far too finicky.

(Dr Inch)

Those women whom these doctors were prepared to operate on had to fit within the parameters of psychological health that the doctors defined. This is interesting given that one rationale they put forward for the work they do has to do with improving the psychological well-being of patients. It is also interesting that one doctor (Dr Inch) who noted that he had 'a very finicky nature' and reasoned that this was why he had decided to work in the area of cosmetic surgery was reluctant to operate on women who he felt were 'far too finicky'.

Doctors accepted that patients that they turned away may well go elsewhere, but there was no system in place to monitor this or to encourage routine psychological assessment before surgery. Dr Mile mentioned that as his practice had grown, he could turn people away that he would have operated on earlier in this career:

> Now I am busy I have the luxury of picking and choosing a bit, so if I twig that someone has an excessive concern about something that is very small, I will follow my instincts and advise them to go somewhere else.

Doctors act as gatekeepers to cosmetic surgery for women through probing the effects of the influences of others on them, their psychological status and whether were deserving of the procedure. (Elix & Lambert, 1999). In so doing they are controlling women's agency and playing a defining role in deciding what it is that is appropriate for women to do with their own bodies.

Doctors and cosmetic surgery

There is no published research that seeks to explain why doctors decide to provide cosmetic surgery procedures in their practices. Given their pivotal role in the industry it is important to understand their motivations and to reflect on how these motivations may impact on the outcomes their patients get from cosmetic surgery. An understanding of what their patients believe motivates these doctors can also contribute to a broader appreciation of the interpersonal dynamics at work between these doctors and their patients.

Women's interpretations of why doctors practise cosmetic surgery

It is clear that more and more women are choosing to have cosmetic surgery and the print and electronic media continues to publicise the practise through articles and programs. Understanding why doctors carry out cosmetic surgery also contributes to our understanding of the industry, the driving forces behind it and the outcomes women can expect from cosmetic procedures.

When asked why they thought doctors practised cosmetic surgery, the majority of women mentioned the economic benefit of the work or a similar answer linked to the profits that could be made:

> I'd have to say a lot of them are interested in the money.
>
> (Ella)

> Money.
> (Fiona)

> To make money. Another payment on the Ferrari.
>
> (Patsy)

> Well its very lucrative, it does cost a lot of money and I really can't see why it should cost so much, compared to other surgery that may be life-saving.
>
> (Ruth)

> I guess they just want to specialize in something and it pays for the holidays and the girlfriends and so on.
>
> (Thelma)

> I suppose it probably started off with doing reconstruction and stuff and I suppose most of them do it for their income.
>
> (Molly)

However, some commented that doctors were concerned with helping women by improving their health:

> I think some do it because they're really seeing that it bothers people and they do worry about a woman's perception of themselves and they do see the difference in the person – before and after and they see that as beneficial to a person's health.
>
> (Clare)

> He is interested in helping people. Yes I think he is genuinely interested.
>
> (Kerry)

Others felt doctors were responding to a demand among women for cosmetic surgery:

> There's a lot of reasons. Obviously demand. A lot of women are obsessed, a lot of women view it as an alternative to the fountain of youth.
>
> (Donna)

> Because there is a demand in society particularly from women who are looking to make a change.
>
> (Hannah)

However, one respondent felt that doctors were taking advantage of women's insecurities about their bodies that was generated by media representations of beauty:

> Because all of these advertisements are all young girls and they all have perfect skin, perfect faces, perfect noses and they always have perfect lips. They know that and they reckon that they're cashing in on it, that's why they do it.
>
> (Hope)

Such responses identify cosmetic surgery very much as a commercial enterprise with doctors responding to consumer demand and profiting from the way women's bodies are portrayed and objectified in the popular media.

One woman, however, commented about the creativity and artistry involved in this type of surgery. This theme was reiterated by doctors and will be discussed later in this chapter:

> he's obviously very qualified but he's also very creative. There's an artistic side to it. I don't think it's enough to be technically good to know what you're doing. There's got to be an artistic side to it too.
>
> (Ffleur)

Although other explanations were given for doctors' motivations, for instance job satisfaction:

> I would imagine that they get a great deal of satisfaction out of it and they probably feel good about themselves.
>
> (Grace)

most women clearly regard cosmetic surgery as a business that makes a significant amount of money for those who practise it. They also see that the demand for cosmetic surgery is consumer-driven and that this demand is being met by medical entrepreneurs. This entrepreneurial aspect of the industry, again, sets it apart from conventional medical practice.

The doctors who carry out cosmetic surgery, however, explain their choice of specialism in a different way, with no reference to the business aspect of the industry.

Doctors' explanations of why they practise cosmetic surgery

Those doctors interviewed were asked why they chose to practise cosmetic surgery. Some plastic surgeons saw it as a natural extension of their reconstructive work. Other doctors talked more generally about their enjoyment of surgery and, in particular, the precision that some cosmetic surgical procedures required. Reflecting what one woman had said, there were also comments about the 'artistry' involved in some procedures.

Surgeon as artist

An interest in art and 'the finer things in life' (Dr Lemon) was mentioned by some of the doctors as a reason for practising cosmetic surgery:

> I've always had an interest ... in drawing, in painting – that sort of thing. And cosmetic procedures require a certain requirement for interpretation of appearance, balance and aesthetic sort of (how will I say) – aesthetic considerations come into play ... I find that performing elective procedures, I have a little more autonomy and I'm able to use those aesthetic things that interest me as well that I find appealing what I do.
>
> (Dr Pound)

Here, there is an acknowledgement of a creatively driven interest in cosmetic work where the doctor has a freer rein to interpret what they view as aesthetically pleasing and recreate this in their work on women's bodies. This places the surgeon above a mere technician that translates what a woman wants on her body and promotes him to a creator and sculptor, whose own aesthetic concerns are interpreted through their work.

Surgeon as perfectionist

Others noted that it suited their personality and their need for perfection:

> I was looking for was something that suits my personality and a very finicky nature, a perfectionist type of nature, I couldn't stand things not being perfect and I still can't and that's why I ended up doing plastic surgery.
>
> (Dr Inch)

> I was always a bit dismayed with reconstructive surgery because it was always so imperfect and it was a very poor replication of how it should be and the closer something is to perfect the more imperfect it seems and so I found cosmetic surgery much more intriguing.
>
> (Dr Trim)

Another doctor linked this need for perfection with an artistic temperament:

> I think plastic surgery attracts people who have that sort of personality, they tend to be almost obsessive compulsive type of people, people who are difficult to live with because they are never quite happy with things ... Obviously it is a very visual type of specialty, what you do you see and they tend to be very visual type people. They like art, they like the finer things in life. All that sort of blends in with the persona.
>
> (Dr Lemon)

Fraser (2003b: 135) notes that:

> The focus on art ennobles cosmetic surgery through the historically legitimate aspiration to beauty in its purest 'artistic' form. It works to replace images of operating theatres and surgical instruments with rather more romantic images of art, in which the female body is the artwork in a very traditional sense. Here, mere vanity is not the woman's motive, nor avarice the surgeon's, rather it is the elevation of our culture, through the central symbol of civilisation, 'Great Art'.

While the identification with artistry elevates the surgeon above 'avarice' in his own mind, it also suggests the surgeon's role in creating the visual outcome of a cosmetic procedure. He is not a mere technician but a sculptor of the female form, an artist who 'utilises and solidifies the objectifying male gaze and the passive female object (Fraser, 2003b: op cit). In this way medical authority and the 'artistry' of the surgeon enable the inscription of his view of 'normal' and 'natural' onto the bodies of women to his satisfaction:

> I'm an artisan if you like. I mean I'm just capable of doing things with my hands and performing surgery and the end result is something that pleases me. I get satisfaction out of that – whatever it is – whatever form of surgery.
>
> (Dr Almond)

While women were overwhelmingly convinced that doctors were motivated to work in this area by financial considerations, only one doctor commented that cosmetic work was more lucrative than other forms of plastic surgery:

> I wanted to be a plastic surgeon and plastic surgeons do cosmetic surgery and as you get older you tend to do less emergency surgery and more cosmetic because it is within hours and it is easier to do it and probably because it is financially more rewarding.
>
> (Dr White)

Through their explanations of why they believe women want cosmetic surgery and their explanations of why they choose to work in this area doctors offer, on the one hand, an opinion that they are providing for the well-being of patients while, on the other, following their own creative urges and what they see as fitting their personality. Thus, they see their work as both beneficial for the patient's emotional well-being and for their own.

Conclusion

This chapter has reported on what motivates women to undergo cosmetic surgery and on why doctors practise it. Women's motivations were reported as internally generated but with external influences. That is, while no direct pressure was put on women by any other person to have cosmetic surgery, there were external forces at play. Women's reports of being teased by others when they were younger and of feeling that they stood out for the wrong reasons attests to the influence of peer pressure as a significant factor in the way they experienced their bodies. The views they articulated about how they needed to change their body to fit into a society that applauds bodies that conform to certain accepted models indicates the extent to which such conformity is promoted and rewarded. Some women felt they could not compete successfully in the workplace when their bodies showed signs of ageing. Others could not come to terms with bodily changes imposed on them through their role as child bearers and mothers. They needed to change their bodies in order to feel at ease in society and not stand out in it.

Having analysed why doctors carry out cosmetic surgery, it is clear that their explanations are different from those proffered by the women. Women see cosmetic surgery as a lucrative business for doctors, while doctors see themselves as artists and creative practitioners who provide women with a psychological service through surgical intervention. The explanation of cosmetic surgery as 'psychology with a scalpel' may have some basis, but doctors do not systematically include psychological assessment as part of their clinical examination. Even when they feel that patients have significant psychological problems that affect the way they see their body, they do not automatically refer on to psychologists. They are most likely to refuse to operate on these women while accepting that they may well seek the procedures elsewhere. They are also careful to ensure that women's motivations for wanting cosmetic surgery are self-generated. However, this self-determination is confined to why they are seeking the procedure rather than to what they want from the procedure. If the doctor views a woman's demands as 'unrealistic', then he will be reluctant to operate. This view of what is realistic is subjective and is embedded in what the doctor regards as being legitimate. Self-determination for women is only allowed within the parameters that are defined by the doctor.

Such strictures are inconsistent with the liberal and individual choice discourse that is characteristic of the means through which women select practitioners. Women's means of selecting practitioners are different from the usual specialist-referral pathway for medical and surgical procedures. Only a minority of women in this study sought referrals through the usual route, their family practitioner. Most gleaned information from friends and media articles. This means that women are exposed to views and opinions about cosmetic surgery that are heavily influenced by texts and advertising strategies. That is, cosmetic surgery is a commercially available product obtainable directly by all who wish to access it. This, together with a woman's drive to normalise that part of her body that causes such anxiety, and the secrecy surrounding her quest, makes her vulnerable to choosing practitioners on scant infor-mation. While the pathway to cosmetic surgery may begin with an active resistance to an unwanted bodily feature, it is quickly beset by the external influences of media representations and practitioner's pro-motion of the practice. When the doctor's clinic is reached, the agency that leads women to cosmetic intervention is further mediated by the tangible power of the doctor to decide whether the reasons women seek surgery are acceptable or not and whether they will operate. If they are prepared to proceed, then the next stage of the process, negotiating the preferred outcome, can be entered into.

7
Communication and Cosmetic Surgery

Regardless of technical ability, someone who appears cold, arrogant, or insensitive is far more likely to be sued than one who relates on a 'human' level. Obviously, a personality that is warm, sensitive, naturally caring, coupled with a well-developed sense of humor and cordial attitude, is much less likely to be the target of a malpractice claim. The ability to communicate clearly and well is probably the most outstanding characteristic of the claims-free surgeon. Communication is the sine qua non of building a doctor-patient relationship.

—(Gorney & Martello, 1999: 37)

This excerpt refers to the important role that communication plays in case of malpractice claims brought against plastic surgeons. Evidence exists to show that communication issues are central to claims handled by medical insurers and statutory bodies around the world that deal with medical malpractice issues (Levinson et al., 1997; Richards, 1990). Even though good communication is central to the any doctor–patient relationship, (Ong et al., 1995) little is known about how this process works in the cosmetic surgery setting. While the role of effective communication has been underlined in primary care, little evidence exists from the tertiary-care setting. As part of the process of understanding the doctor–patient encounter in cosmetic surgery, one of the key themes this study examined was communication between women and the cosmetic surgery practice.

This chapter sets out what constitutes communication in the cosmetic surgery setting. The chapter presents communication in a broad context. It reports what patients and doctors say about communication and communication tools in cosmetic surgery. This includes the ways in which doctors communicate with potential patients, how they negotiate the aesthetic outcome of cosmetic surgery and what communication tools are used in the clinic. It also describes the ways in which women explain what they want from cosmetic surgery. Women's reports of communication in the clinic with doctors and practice staff are also explained.

Communicating with potential patients

Doctors communicate with potential patients in a number of ways. While a small minority of doctors reported that they did not advertise their services, most did advertise through some means. This is not surprising given that cosmetic surgery is a product that doctors offer to the market. Unlike other forms of surgery, it is not usually carried out for functional or health reasons and it does not fall within the funding parameters of the public health budget or private health funds. Therefore, doctors need to attract their clientele through one means or another.

In this study, a number of advertising strategies were mentioned. For instance:

> Mainly by word of mouth. Now we do have a listing in the Yellow Pages and we have a listing with the ASPS and we have brochures available. But most referrals come from patients that you have dealt with who bring their friends along.
>
> (Dr Inch)

> Our work comes from a number of areas, a combination of word of mouth, radio and Yellow Pages.
>
> (Dr Mile)

> We have a newsletter that we send out to our patients telling them of what is available here. We have information evenings which we advertise externally. We put a notice in the newspaper or a magazine such as 'Insight' in Melbourne and that sort of thing, and we invite people to come along for a brief talk and I explain about certain procedures. What is available, what is involved and I answer questions. We also advertise in the Yellow Pages and we occasionally put advertisements in newspapers if there is perhaps a new technique that becomes available.
>
> (Dr Pound)

Direct contact with family practitioners and the beauty industry were also mentioned as strategies for advertising:

> I guess we advertise our wares by being didactic and educational to general practitioners and to peers and writing papers ... I used to advertise in newspapers, even on TV. It is advertised in the country and one of our nurses had to do the rounds with us and connect with local doctors and beauticians and hairdressers and the like.
>
> (Dr Cherry)

Doctors also commented that advertising had become necessary for them because other practitioners were advertising and they needed to compete:

> so you have to market just to combat other peoples marketing.
>
> (Dr Yard)

> Currently I have a notice in the Yellow Pages. This is a new development because I haven't advertised before. But I think cosmetics have moved in Melbourne in the last few years and it is harder unless you advertise.
>
> (Dr Almond)

Other specialist medical practitioners list in the Yellow Pages but are very unlikely to advertise elsewhere as they rely on referrals from family practitioners. Indeed, in Australia as in many other parts of the world, their services usually attract a rebate through the public health budget and/or private health insurance, only if there is a referral. Cosmetic surgery, however, does not attract this rebate, unless the surgery is carried out on medical grounds, for example, breast reduction. As such, the industry needs to be able to access patients through other means.

Given that visual, audio and print media permeate our lives on a daily basis, these avenues are the obvious choice for doctors who wish to advertise their services. Any perusal of adverts for cosmetic surgery shows that the target market is women. The most common print adverts in magazines and newspapers being for liposuction and facial surgery. For example one practitioner, in an advertisement for facial surgery which had a drawing of a smiling woman, asks 'Why not discuss the way you look with someone who really knows?' This implies that the doctor is the expert in female beauty and can provide 'solutions to your particular concern'. The advert goes on to say that 'It's not such a drama as people think'. Here women are led to believe that they can access facial surgery through a sympathetic and knowledgeable doctor who will understand what *they* want and can respond to their individual needs. They are also reassured that the process of changing the way they look is straightforward. Such advertisements lead women into a process that is far from simple or unproblematic and belie the reality that cosmetic surgery is real surgery with real risks that may potentially result in unpredictable and unwanted outcomes.

How do women find out about and choose practitioners?

The women in this study used a number of sources of information and referral to access their practitioner. They received information about cosmetic surgery and practitioners from a range of places, including friends, magazines and newspapers, the Yellow Pages and family doctors.

The media

> There's a lot of stuff in the media and magazines about cosmetic surgery, it's quite an acceptable part of today's community and lifestyle.
>
> (Hannah)

Coverage of cosmetic surgery is seen often enough in magazines and, in particular, in women's magazines for it to be expected and accepted in the print media. Such media coverage seems also to give some women the confidence to seek out procedures and the practitioners who carry them out. It seems that magazine articles that mention practitioners are successful advertising tools in attracting women to those practitioners' practices:

> I read in magazines and actually it was in the *Women's Weekly* when Mimi had her operation and you know she looked fantastic. I saw the surgeon that she had do the operation and I was just looking through the Yellow Pages phone book one day and found that I could just ring him up. I thought he did a good job on her and I must admit that I didn't go round and see any other surgeons, which probably wasn't the right way to go about it.
>
> (Irene)

While doctors are not paying for advertising through this medium, their status is enhanced and the procedures they carry out promoted through media coverage. Such media exposure also encourages women to have confidence in their abilities:

> I'd heard about him on the radio or television. That's the cranio-facial chap. I thought that if he deals in the head and face then he would be more relevant to me. So I had a little bit of confidence in that because that was his specialty. The other cosmetic surgeon, he had quite a few articles in the paper, he's been on television, he was with another chap and both of them were both quite involved in laser surgery.
>
> (Megan)

The use of the media also acts as a catalyst for women to take action and make the decision to undergo a procedure that they may have been considering for a while:

> The breast implants I heard about on the radio, it was an advert, I always knew you could get it done but that's just what prompted me.
>
> (Tanya)

Interestingly, of the seven women interviewed who had complained formally about their treatment, five had accessed information about their doctor from the media:

> In the newspaper.
> (Diane)

> Through magazines and friends and such like.
> (Dora)

> Unfortunately, I didn't do too much research, only articles and TV programs. I got caught up in the advertising really for this particular person I went to, and I only went to him.
> (Freya)

> Through television, and an article in *New Idea*.
> (Hope)

> I got a referral to this doctor ... who is often on TV current affairs shows and in magazines.
> (Lin)

It is not possible to say whether accessing information about cosmetic procedures through the media contributes to negative outcomes in some way. However, the nature of articles in these media may lead women to expect outcomes that are less painful and invasive than they actually are and to believe that the doctor can give them a guaranteed result. Doctors who utilise the media to promote their practice and attract patients may be reinforcing such attitudes through the information they provide in their advertising material and in the magazine articles that they cooperate in producing.

Friends and other personal contacts

Other women followed the recommendation of friends, family and other personal contacts:

> Through talking to people, clients and people like that.
>
> (Chloe)

> One of my other friends had a breast enlargement and she was happy with him and I went to him for a consultation and 12 or 13 months later I followed through and had the procedure.
>
> (Greta)

> through word of mouth I had heard about the practitioner who was meant to do the procedure.
>
> (Lorna)

> Well my sister was a nurse and she suggested the plastic surgeon. That was for the first one and I used him for the second procedure as well.
>
> (Ruth)

> The surgeon I used was highly recommended by a colleague.
>
> (Sonia)

> Through a nursing friend, actually two friends who worked in cosmetic surgery clinics, and they recommended the same doctor.
>
> (Thelma)

Family practitioner referrals

Referrals through their family practitioner were not as common as other sources of information for women. This may be because women are reluctant to disclose their wish for cosmetic surgery to their practitioner or because cosmetic surgery is seen as a form of specialist medicine that can be directly accessed without a family practitioner referral. However, some women did seek a referral from their practitioner:

> Probably through the media and I spoke to my family practitioner about it.
>
> (Grace)

> My family practitioner recommended the one I saw recently and her name appeared on the list of the Australian Society of Plastic Surgeons.
>
> (Ruth)

From the perspectives of the doctors interviewed, there was no general consensus about the need for a practitioner referral. One of the doctors interviewed felt a family practitioner referral was preferable for two reasons, one as a screening tool for those 'not suitable' for cosmetic surgery and secondly from the perspective of their overall health:

> We suggest to people when they ring up to make a booking that they see their general practitioner first. I think that it is a valuable screening process. I mean there are some people who just aren't suitable for surgery or for cosmetic procedures and I think that it is important that they can be screened first and save them the trouble and effort of coming all the way to see me. It also keeps their family doctor informed of what those people have in mind and I think that is pretty important from the point of view of just general healthcare.
>
> (Dr Pound)

Some doctors felt that there was hostility towards cosmetic surgery from the medical profession that could lead family practitioners to discourage patients from undergoing procedures:

> There is a lot of hostility in the medical profession about this work.
>
> (Dr Black)

> A number of people who want to come and see me are sometimes discouraged by their general practitioner, which is based on ignorance, possibly jealousy, envy – all sorts of things like that.
>
> (Dr Step)

There were reports of family practitioners refusing to provide a referral for cosmetic procedures, but there were other reports that family practitioners were becoming more accepting:

> it's been interesting over the years how many doctors refused to provide referrals. Years ago some family practitioners would say 'I won't give you a referral' and even now some patients have to shop around.
>
> (Dr Trim)

> Oh, I think that at the family practitioner level there is a lot more acceptance. In fact, I get quite a few family practitioners come in for treatments themselves.
>
> (Dr Green)

Of the women who specified the number of doctors they had consulted with before having their procedure, 17 had consulted with one doctor only, five had consulted with two doctors, three had consulted with three doctors and two with four or more doctors.

Print and electronic media advertising and personal recommendation were the most common source of information about practitioners for women in this study, a fact that supports the contention that cosmetic surgery sits outside the boundaries of conventional medicine and the usual rules of referral from a family practitioner. Only one of the women who complained about her treatment had been referred by a family practitioner. All but one of the other complainants had accessed their practitioner through print or electronic advertising. The coverage of cosmetic surgery in the media normalises what are invasive procedures, making them seem less traumatic than they actually are, and doctors who use the media to promote their work may be complicit in promulgating such an impression.

Communication in the clinic: The women's perspectives

The first face-to-face contact women have with cosmetic surgery services is when they attend the clinic. Here begins the next stage of the journey towards cosmetic surgery and their engagement with doctors and staff in the practice.

First impressions

When a woman has decided which doctor or doctors to approach, the next step is to have a consultation. Many cosmetic surgery practitioners practise from rooms that are expensively decorated and aesthetically pleasing, giving the impression of affluence and offering a high degree of comfort. They often have plaques detailing the qualifications of the practitioner displayed on the walls. This can communicate to patients that the surgeon is successful in his or her practice and that they are well qualified in the work they do:

> He certainly wasn't very good on the patient relationship side but the surgery was nice and I thought that the presentation was good and I thought that if this is the way they operate then he must be good. He had the right things on the wall.
>
> (Freya)

Having 'the right things on the wall' indicated to women that the doctor was highly qualified in their field. However, while many doctors have significant postgraduate qualifications, some certificates may refer to short courses undertaken rather than to advanced degrees or diplomas. Many of these short courses can be undertaken in the US where some Australian cosmetic surgeons go to learn new techniques or learn to use the latest instruments.

Many practices also make available leaflets that outline the processes involved in various surgical and non-surgical cosmetic procedures. These leaflets may be produced by the practice, by a professional organisation or by pharmaceutical or other companies that supply the cosmetic surgery industry:

> He had a brochure in his surgery when I was in waiting room with all the different cosmetic procedures and I just picked it up.
>
> (Clare)

Some practices also have leaflets from companies who offer loans to patients to cover the cost of their surgery. Such a practice may encourage patients to go ahead with a procedure that they cannot otherwise afford at that time or undergo additional procedures they may not have planned to have. For instance, Lin who was seeing the doctor about her eyes, had not intended having her nose operated on but decided to go ahead when she saw that she could access a loan to cover the cost:

> Anyway he asked me what I wanted to do about the nose, he said, 'Do you know what we say in my profession, never waste an anaesthetic'. It was a couple of thousand dollars but I thought what the

hell, because they had these leaflets in the surgery with information about loans that you could pay off after the surgery.

Such leaflets act to promote cosmetic surgery by outlining both what can be done and how it can be paid for. Such a practice places cosmetic surgery firmly within the commercial market in the same way as a department store that offers loans for the purchase of costly items. While doctors are not involved in providing loans, they do benefit from the promotion of such a service to their patients.

The surroundings that doctors consult in and the type and range of written materials they offer in the waiting room influences what a woman thinks about those doctors, their success, their qualifications and whether they are able to provide the outcome a woman expects. These things act as part of the communication toolbox that doctors use to interact with patients when they come to the clinic. They successfully promote the range of procedures that doctors provide and act to encourage women to undergo those procedures.

Practice staff

Most practitioners employ support staff to assist them in dealing with patients. These staff can be an important bridge between the doctor and patient and can fulfil a number of roles. These roles include carrying out computer imaging, ensuring that patients complete documentation relating to the procedure they are about to undergo (including informed consent for procedures), handling enquiries relating to procedures and dealing with appointments and payment. For some women the contact with practice staff was a positive experience and was an important factor in determining whether they would make an appointment with a particular doctor or not:

His staff were particularly friendly, I felt comfortable with them. I then did subsequent phone inquiries with other practitioners and just again it all depended on how their staff reacted and handled phone inquiries.

(Donna)

Staff also had a reassuring role for women considering surgery:

I spoke to his secretary who assured me how good he was. She'd had things done and it wasn't painful.

(Hope)

While women appreciated some intervention by practice staff, they were not happy when staff were used to substitute for the doctor who was too busy, or unwilling, to see a patient. Such substitution with staff that cannot provide patients with adequate information can lead to patient dissatisfaction and frustration about not being able to have their questions answered:

> Well, I went back and forth to see her a few times but couldn't get to see her. I only saw her staff and really they just couldn't tell me what I wanted to know.
>
> (Dora)

Staff who are under pressure and who are expected to act as gatekeepers to the doctor also contribute to patient dissatisfaction:

> Every time I went in there the nurses seemed to be under a lot of pressure and they seemed confused and all I wanted was to see the doctor but she was like some kind of goddess. Unless I was dying or something I wasn't going to see her.
>
> (Lin)

Practice staff were often the first contact point for patients who were anxious about the procedure they were about to undergo. Usually it is the role of the doctor to discuss such issues with patients, but practice staff were often used as substitutes for the doctor. The following quotes show that this was not always successful:

> But as it was getting closer to surgery I was getting more and more anxious and that's why I was ringing him and his receptionist was really rude.
>
> (Lorna)

Here the receptionist acts as gatekeeper to the doctor, a role many medical receptionists undertake. However, the doctor had been unwilling to answer the patient's questions in the consultation and the receptionist was unable to:

> But every time I asked him questions about what size bra I would go up to so that I knew if I was getting implants that were big enough … he'd umm and aah and skip over it.
>
> (Lorna)

Even when the staff were seen as helpful, it was acknowledged that some patient questions could only be responded to by the doctor:

> But I was still not happy really so I spoke to one of the staff on the phone and she was very helpful but really she couldn't tell me everything I wanted to know, it should have been the doctor who told me that.
>
> (Dora)

The responsibilities and qualifications of practice staff varied between practices. Some of the staff were trained nurses who were able to discuss medical details of procedures with patients, while others were there to deal with administrative and financial matters. These roles sometimes seemed to be blurred and it was not clear that women understood what each staff member was responsible for, or that staff were qualified to carry out the responsibilities allocated to them. In the following case the staff member was responsible for getting the patient to sign the papers that detailed consent to the surgery, even though the patient was not sure what she was agreeing to:

> I hesitated at one of the papers (contracts). I wasn't going to sign it. My husband said, 'What's wrong?' And I said, 'I don't like all this. I don't know what is going to happen to me because nobody has explained anything to me'. The assistant's words were 'this is a normal procedure. Doctor has done so many operations, like you, but you understand that the law requires us to have these papers'. I wasn't comfortable.
>
> (Diane)

The role of the staff member here was to obtain consent regardless of whether the patient had adequate information. Doctors verified that staff were used for such a purpose but their perspective about how the process was undertaken was different to that of the women quoted above:

> I speak with patients but then I send them to one of my cosmetic nurses who has a form and asks the patient to sign each point as they've gone through it. The nurses will spend at least an hour with the patient going through all the side effects.
>
> (Dr Yard)

> they have a pre-operative appointment with M (a staff member) ... and virtually every risk is pointed out to them.
>
> (Dr Metre)

These quotes support the evidence from women that practice staff were often given the responsibility of achieving the consent of the patient. This can, of course, be problematic if women are poorly informed about exactly what they are consenting to and when practice staff do not have the knowledge or expertise to provide adequate information. This will be discussed more fully in Chapter 8.

Deliberations about the outcome of surgery

A crucial, but difficult, part of the cosmetic surgery process is to achieve a shared understanding between doctor and patient about the outcome of the surgery. Communicating what one wants from cosmetic surgery is difficult for women and, as we will see later in Chapter 8, having a less than satisfactory aesthetic outcome is one of the risks of undergoing such surgery. There are a number of strategies employed by doctors to demonstrate what outcomes can be achieved from a procedure. These include the use of before and after photographs of past patients and computer imaging. These tools are used in facial surgery and liposuction.

For women who had breast implants, the approach was different. It included negotiating what size implant would suit the woman's build and, for some, experimenting with implants placed in their bra before deciding on which size they wanted. Despite these strategies all but one of the women who had breast implants said that they wanted a C cup, but all ended up with breasts that were larger than they had expected. This will be discussed later in this chapter.

Photographs of past patients were used to illustrate the sort of changes that could be achieved through a cosmetic procedure. These photographs were either shown by the doctor or by practice staff. Some were shown as examples of what could be achieved. None of the women mentioned that she had been shown any cosmetic surgery failures. As one woman noted:

> Yes, he showed me heaps. You could see the difference. But then again if he'd had any failures he wouldn't have them in there.
>
> (Molly)

Only one woman reported that she had been shown photos that showed the recovery process after surgery and this gave her an understanding of how her own recovery would proceed:

> Yes, he showed them to me to make me understand what would happen. A lot of people have this perception in their head that you

just walk in and a couple of hours later you walk out. He showed me
photos of a woman taken daily to show what happened.

(Ffleur)

Other doctors sketched with pen or pencil what the outcome might be
and did not show any photographs. Even though Greta had no clear
picture of what the outcome might be, the discussion that had taken
place was reassuring enough for her to 'be happy' with what the doctor
was suggesting:

> We discussed it for quiet a long period. He didn't show any photos
> he just did a sketch, put it up to my face and looked at it through
> the mirror, but it's very hard to see what it's going to be like, but I
> was happy with it.

Although doctors were not asked directly if they used before and after
photographs when consulting with patients, a number of them mentioned that they did and viewed this as a good strategy for negotiating
an outcome with patients.

One doctor saw before and after photos as a good way of explaining
what could go wrong in the procedure:

> The next stage is to look at what can be expected after the surgery
> and photos of before and after results are useful. I show them what
> can be expected of say a breakdown of a wound after breast surgery.
> I show them good results and I show them bad results so that they
> know that they can get a bad result. Here is another one of a breast
> augmentation and it's a lousy result and I show it to them so that
> they know that they can get a bad result.

(Dr Inch)

The use of computer imaging

Computer imaging is used by doctors to provide patients with computer
generated images of the outcomes possible from surgery. While this may
have the potential for developing an agreed outcome of surgery, and in
particular facial surgery, it only shows what may be possible rather than
what is actually achievable. It may represent an image of what women
may want from their surgery but not necessarily what a surgeon can
achieve. This difference between vision and reality and the efficacy of
computer imaging is something we know little about. Despite this it is

widely used as a tool in communicating what is possible in cosmetic surgery.

For the women in this study, imaging was carried out either by the doctor or by practice staff:

> he had a computer ... he put my photo on it. You know, he showed me somehow. He said, 'I must show you the difference. It will just be with the bump gone, you won't have any difference in the length or the width or anything. It will just be with the bump gone'.
>
> (Clare)

> He had one of those computer modelling systems, he took some photos and downloaded them onto the computer, and then one half of the screen had the photos and on the other half he digitally adjusted my photo to show how the nose would be after the surgery.
>
> (Hannah)

> Anyway the consultation with the doctor ended and then I went to see a nurse and she had this computer imagery that we played around with and in my head I looked a lot different on the computer.
>
> (Lin)

Computer imaging was both confronting and comforting for some women. It brought into sharp focus that feature that they so disliked while, at the same time, it gave them hope that it could be changed:

> I was so shocked about my profile and I was a little dismayed about how bad my profile really was. With the new profile it was like a flood of relief that something could be done.
>
> (Hannah)

Women seemed to focus more on getting rid of what feature was bothering them, rather than on what the new feature would look like. However, one woman reported that she had decided not to have the nose outlined for her in computer imagery, but her doctor took no notice of her decision:

> I went to get the imagery to find out what the nose would be like if it was bigger, because they were going to build the bridge up bigger and on the computer that looked OK. I didn't notice curvatures or

anything. Just before the operation I said that I wanted a straight
nose and he said 'No, no this one would suit you better'.

(Chloe)

Charlotte reported that her doctor was somewhat sceptical about computer imagery:

The other doctors, they took a photo of you as you were and then
did computer imaging of what you would look like afterwards. We
took those photos to the public doctor and he basically said that was
rubbish. There was no way that they could tell us what the outcome
would be. He said, 'Don't think that this is how it's going to come
out because that is not necessarily what is going to happen'.

And Diane felt that computer imaging had given her an unrealistic
expectation of what she would look like after the surgery:

The computer imaging has to stop. The computer image changes
everything. And it's not true because it's a computer image! I realize
that now but at the time, the way they did it and the way they joked
about it – they make you think, 'oh my God, how good I'm going to
look! And I'll feel so good!' I wanted to look exactly how the computer said I was going to look.

It is clear from the responses of these women that there are mixed views
about the efficacy of computer imagery in cosmetic surgery. It helped
some women understand the changes that could be made, but left others feeling that they had not got what they had seen on the computer
screen.

Breast implants

The process of negotiating the size and type of breast implant was
slightly different than that of other procedures. Some women who had
breast implants reported being shown before and after photos, but none
mentioned having computer imaging. For most women, negotiating
the size of implants began with them describing to their doctor what
size cup they wanted. Some doctors then asked women to place something in a bra that would replicate the breast size they wanted:

What I do is I tell them to go and get the bra size that I think I can make them and you put cotton wool into it (you fill it up not too tight). They put a T-shirt on and have a look at themselves in the mirror and that will give them a very good idea with clothes on what I can do.

(Dr White)

Only Lorna wanted breast implants that were larger than a 'C' cup and she wanted 'big and obscene'. Every other woman just wanted to have a 'C' cup. In negotiating the size of implant, the doctor would refer to the woman's build and what they thought was most suitable. However, whatever their build was, women reported that the size of the final implant was larger than they had wanted or expected. It was very difficult for women to imagine what their implants would look and feel like. Nevertheless they wanted implants that were not too large. As Thelma said:

I didn't want to be Pamela Anderson.

Tanya 'had no idea' about how she would look after the surgery and followed the advice of the doctor:

I would have liked to have seen what it was going to look like first, but I had no idea. He just said choose a size, he said that 'C' cup's the most common.

When the implants were ordered, she was unsure as to whether they would be too big. Sensing her uncertainty the doctor's secretary had ordered smaller implants as well but on the day of the operation:

the doctor kind of talked me out of it, but now I wish I had gone a smaller size.

The doctor did not seem to sense Tanya's ambivalence about the size of the implants, but went ahead and used the ones he thought was best. This is a recurring theme among women who had breast implants. Ultimately they got the implant of a size that the doctor thought best and not what they wanted. This lack of knowledge about the outcome was not addressed in a positive way by doctors before the surgery. Afterwards, as long as there were no post-operative problems, they were not interested in listening to women's feelings about their new breasts:

> They are bigger than I expected them to be and I did mention that at my first check up with him, but he's not interested in hearing that. As far as he was concerned I looked fantastic and there was no infection. I'm a D but I would have preferred to be a C.
>
> (Thelma)

Here, the doctor's opinion of how Thelma's body should look overrode the fact that she thought the breast implants were too big. She reported that:

> I've had a lot of negative response from females. Also I have a lot of men staring at me in the street. I feel naked. They are like headlights I believe.

A more extreme example of a doctor making the decision on the size of implants occurred when Patsy woke from her surgery to find her new breasts much larger than she had asked for:

> It started when I woke up and I had these enormous breasts and the surgeon just said to me 'I made them a bit larger, because you'd only want them larger a year down the track'. They were a double D and I was supposed to be a C cup.

She felt angry at her surgeon for imposing the breasts he wanted on her:

> obviously he hadn't listened to me. I wanted to be C cup like I was before not a double D.

Patsy had felt uneasy about his method of showing different breast sizes to her before the surgery:

> In hindsight I shouldn't have used him because he showed me these pictures of breasts from *Penthouse* magazines partly covered up, to show what he could do.

For Sonia, the doctor had suggested the size of implant but said he could not guarantee the outcome and she was happy with this. It is significant here that her doctor spent as much time as she required with her and had not guaranteed the outcome of the surgery:

He outlined what he thought would suit my body shape and my height and he recommended the size, but he couldn't guarantee the size, what I got was what I got. I was happy with that. I looked at lots of literature, asked more and more questions. I felt that I could spend as much time as I wanted to with him.

For Tanya and Lorna this was not the case:

I don't feel he really listened to me, it was more 'OK, let's get this over and done with'.

(Tanya)

Lorna contrasts the approach of two doctors she spoke to and said:

I felt rushed. He got me in there and he had another appointment and he showed me the photos, these are the implants, these are what you can have, see you later. With the second one, even though he had people in the waiting room, he didn't finish the appointment until I was ready to finish the appointment.

She went on to say:

Look for a doctor that wants to spend time with you, that will answer all your questions, a doctor that doesn't brush you off. If a doctor doesn't want to answer your questions just give him a miss. Once you've made that decision, it is something that you have got to live with forever, so spend time choosing your surgeon. Don't get excited because your bank loan has come through or you've saved enough money to have it done. Get the excitement factor out of the way.

This encapsulates the view of many women interviewed, that effective communication with the doctor depended on time and their ability to provide adequate and pertinent information. In the end, it was the effectiveness of the communication between doctor and patient, not between any other member of the practice staff and the patient, which was a pivotal part of negotiating a satisfactory outcome.

The consultation

Cosmetic procedures do not begin with the scalpel and end with the stitches. As we have seen, for women, effective communication was an important part of the treatment. In particular, the communication that takes place in the consultation with the doctor. For the patient, the goal of the consultation is to achieve the outcome they desire. Although women who seek cosmetic surgery may have informed themselves as best they can about the procedure, they are not experts in this area. But they do understand the process of communication. Ivana recounted her experience of selecting a doctor in this way:

> I had tried about 4 or 5 plastic surgeons. Some of them I felt uncomfortable with and this one I didn't. Well, they weren't prepared to listen to what I had to say and I needed to feel that I had a certain amount of control instead of giving all my trust to this one person. I didn't feel I had really gotten my views across and we didn't spend enough time talking about it. Not that there was a breakdown in communication, it's just that the doctors had specific ideas, you had specific ideas and somewhere in the middle they have to meet. I found with a lot of doctors that wasn't the case. Some doctors would even say, 'Have you thought of getting your eyes tidied up while you're here'. I thought that was extremely unprofessional.

Women feeling 'comfortable' or 'uncomfortable' with their doctor was a recurring theme in their narratives about the consultation. Those who felt comfortable with their doctor were more likely to feel that communication was good and that they had achieved a satisfactory outcome:

> Yes I felt comfortable with him. He was very nice ... he had time for me.
>
> (Irene)

> He wasn't pushy, and I felt comfortable with him. I wouldn't have gone ahead with the procedure if I hadn't felt comfortable with him.
>
> (Greta)

Women who had unsatisfactory outcomes, not surprisingly, felt less comfortable with their doctor:

> Looking back. I never really felt very comfortable with this doctor because it was like a process, you were pushed in and out.
>
> (Lin)

There were also reports that some doctors acted in what some women saw as an inappropriate manner. For instance, having telephone conversations about personal matters while consulting, treating women in a flippant manner or being rude:

while I was in there he took a phone call about some piece of art he had in his office and he was talking about this on the phone and then he started telling me that this piece of art could go up in value!

(Lin)

I was really unimpressed with him. He kept me waiting for an hour and then he came in to the room and he said, 'Well my dear' ... without apologising to me ... 'Well my dear, now how can I help you?' And I said, 'Number one I'm not your dear, so you can stop calling me "your dear"' and I said, 'It would have helped me if you hadn't kept me waiting so long. I've been waiting an hour over my appointment time. I was just getting up to leave'. So he got off on the wrong foot with me, and he said, 'Oh I'm terribly sorry'. And I thought, 'you're not a bit sorry, you're just a smarmy bastard'. So that was it for me with him.

(Molly)

He basically was extremely rude over the phone. All I wanted to know was what I should do. I went through a number of phone calls with him that were quite distressing.

(Freya)

It is during the consultation that the doctor has the opportunity to listen to the patient and the patient can explain what she wants from a procedure. From the reports of the women interviewed, it seems that many doctors did not allocate sufficient time to the consultation. According to some women, and in particular those that had complained about their treatment, relatively little time was spent in consultation with the doctor. One woman reported spending as little as seven minutes with the doctor:

I spent seven minutes with the doctor. (The doctor) spent most of the time working out the bill not talking to me.

(Diane)

Others spent slightly longer:

The first consultation with her was about half an hour and then it was just her staff I saw.

(Dora)

It is difficult for doctors and patients to communicate effectively and build rapport in such short consultations, and it is not surprising that this type of consultation in cosmetic surgery seems more likely to result in patients being dissatisfied with the outcome of their surgery.

Those women who had complained about their treatment felt that communication had broken down with their doctor. Some of these women said that they had 'a gut feeling' about the practitioner. That is, they felt some disquiet but couldn't really say what bothered them. Others were surprised at how little time they spent in consultation with their doctor before the surgery. Under these circumstances, there was no time to build a relationship with the doctor, or to adequately outline their needs.

Other women interviewed valued the fact that their doctors were prepared to spend as much time with them as they needed before proceeding with surgery:

> I had three or four consultations and if I needed more then I could have had them. There was no pressure.
>
> (Greta)

> Face-to-face I saw him twice. I went away and thought about it for a couple of months. I guess a half an hour consultation, but I didn't feel that he was rushing me or anything.
>
> (Irene)

There were women who were not dissatisfied with the outcome but felt that the communication during the consultation with their doctor was poor. Hannah said:

> I'm not as satisfied as I could be with my doctor's interpersonal communication skills, but I am happy with his technical skills and the outcome. I feel it's almost by accident that he got it right. He was slightly dismissive on every occasion I spoke to him.

Of course, if Hannah had not been satisfied with the outcome, her discontent with the quality of the communication with the doctor may have had a significant influence on whether she chose to complain about the outcome.

Ivana who was not happy with the outcome of her surgery, but had not complained, said:

Communication was good even though I wasn't satisfied with the outcome. I felt that he spent time with me and is happy to get to where I want to be.

(Ivana)

Satisfactory communication is something that these women felt was very important as part of the process they were going through. Technically good results are not the only way they measure a satisfactory outcome. Communication includes the amount of time the doctor spends with a patient, giving their undivided attention to the patient, behaving in a personable manner, that is, not being arrogant or dismissive of the woman's concerns, empathic listening and understanding the reasons why women decide to have a cosmetic procedure. Patsy spoke of the way her decision to have breast implants was misinterpreted by her doctor:

(The doctor) said, 'They'll be like bees to a honeypot now'. He meant men, and that was never part of any reason why I had done it.

When asked to describe how she felt she had been treated by her doctor she said:

Coldly. Arrogantly. It was like he was turning me into the vision of his perfect woman. He didn't listen to me. He wanted me to look a certain way and that's what he did. He didn't do what I wanted. He did what he wanted.

The view that this doctor wanted Patsy to look a certain way was echoed in various ways in other interviews. Diane commented that when she went for a post-operative visit to her doctor after a lower facelift, she looked at other women in the waiting room and they all had the same 'look' as her. In such situations communicating what *they* wanted from the surgery had not been possible with doctors that were deaf to their needs or whose technical skills could not produce the outcome they wanted.

Some women seemed to 'opt out' of the decision-making process and leave it up to their practitioner to decide the size of implants, the shape of the nose etc. For some, this left them with an outcome that they weren't happy with, but they felt that this was their responsibility as they had left the final decisions up to the doctor. When asked if she felt the doctor understood what she wanted, Thelma said:

> Definitely not, because I'm bigger than I expected to be. I really left it in his hands, I left it to his discretion. I told him I wanted to be balanced, but I didn't want to be Pamela Anderson. All I said to him was that I wanted to be back to what I was before, to fill what was there.

Hannah was happy enough with the outcome of her surgery, even though she had left the decision making to her doctor:

> I left it pretty much to him. I decided to follow his recommendation and have the bump taken down and the nose shortened and at a later date he and I would decide if the implant was necessary.

Both these examples suggest that, in leaving the decision making about the outcome of cosmetic surgery to the doctor, women become passive recipients rather than active decision makers. These women acceded to the doctor's knowledge and opinion about the aesthetics of their body.

Knowledge gained so far about the way professionals communicate shows that too many recede back to technical skills and forego communication when negotiating outcomes. Implicit in this is that the professional has the upper hand through their superior knowledge and skills. The problem here is that doctors who revert to this behaviour often unable to communicate effectively with the patient to ascertain what the patient wants. Aesthetic surgery requires a shared understanding of outcome, benefit and risk. The way women measure the success of an aesthetic procedure is not only through the technical outcome. They take into account the nature and quality of the communication they have with the doctor.

Communicating with patients: The doctors' perspectives

Doctors' reports on their communication with patients revolved around negotiating the nature and extent of the surgery and whether the doctor thought the surgery was necessary. No direct questions were asked of doctors about the amount of time they spent with patients or the number of consultations they had with patients before or after surgery. Dr Step did comment that when he first started to practise, he felt he made the mistake of operating on a woman he had consulted with only once. The outcome was not good and the patient was dissatisfied:

Normally I see people twice as a rule (it is now an iron rule), but back then, this was in my earlier days, she said, 'Look I hate my breasts, they're terrible, they're disgusting, they're revolting' – which should have been a tip-off as well but you learn ... 'And anything you do, you can't make them worse'. So I did her after one consultation which was a mistake.

It was clear from interviews with doctors that they felt a substantial amount of consultation time was spent in discussion about the nature of the surgery that women were seeking, and what were the best options for women to achieve the outcome they wanted. This is, of course, in contrast to what many women said about their consultations with doctors.

Communication tools

Doctors reported that they used a variety of tools to help them and their patients to achieve a shared understanding of what the outcome of surgery would be. The tools they reported using were photographs, computer imaging and pen and paper.

Some of the doctors were sceptical about computer imaging and did not feel that it was any better than photographs or drawing:

> Now, I mean, I know there are some people who use computer imaging and all the rest of it. I don't – I just use my Polaroid camera.
>
> (Dr Sand)

> computer imaging. I don't do that because you know, in my mind, you can do what you like with a computer, because the computer is very functional. I think you can just draw things.
>
> (Dr Salt)

While others felt that the process was very useful:

> I use computer imaging to show people, particularly with nose jobs, it's very helpful.
>
> (Dr Mile)

Others used it routinely to detail what women wanted from their surgery and what was possible. Computer imaging was also used to show whether the outcome that was possible was worth the investment or whether the surgeon could achieve an outcome that a patient wanted:

It comes out usually very clearly in the computer imaging. It might give them an idea that they don't need the procedure or they are not really going to benefit very much from it.

(Dr Metre)

Computer imaging was seen as a useful tool for dissuading patients with 'unrealistic expectations' by one doctor:

You can do computer imaging in certain circumstances and I have found that to be very helpful ... Computer imaging often exposes the person who is unrealistic because they persist in telling me I can do it and I tell them I can't.

(Dr Lemon)

Strategies for measuring improvement

Explaining the likely benefit of cosmetic surgery was discussed by doctors. There were doctors who reported that when communicating the benefits possible from cosmetic surgery, they talked with patients in percentage terms:

I always quote 70 per cent improvement.
(Dr Green)

what I say to patients is typically what I will say in the conversation is, 'Look, although I will aim for 100 per cent result with your facelift, or eyelift or nose or breast or whatever, the reality is that I may only get 90 per cent, or I may only get 85 per cent'..

(Dr Dance)

This was seen as a way of explaining to patients what was possible, curbing any 'unrealistic' expectations they may have and balancing benefits against the cost. As Dr Trim put it:

They come in and ask 'Can you fix my nose?" or 'Can you do my eyes?' and there is no advantage to doing that – this is only one or two per cent.

Women also reported that doctors used a variety of strategies for explaining what could be achieved from surgery:

The practitioner I went to said very clearly 'look you're not going to – I tell all these clients – you're not going to end up looking like Elle MacPherson. It's more of a re-shaping'.

(Donna)

As mentioned in Chapter 6, doctors noted that the negotiation around the expectations women might have of their surgery was an important part of consultation and of the communication between doctor and patient. The assessment by doctors that women's expectation of outcome was 'realistic' was a constant theme in the interviews with doctors and usually related to patients having too high an expectation of the outcome of surgery. Dr Dance explains under what circumstances he refuses to operate on women:

I think if the procedure won't give the result the patient desires or requires and I figure that if I feel the patient has too high expectations and is not realistic about understanding what is involved in the surgery.

The issue of 'unrealistic expectations' will also be discussed in Chapter 8.

If negotiations about 'realistic expectations' fail then doctors will refuse to proceed. There are other reasons why doctors refuse to carry out procedures, which include the risks outweighing the benefits and the doctors' assessment that women are doing it for the wrong reason:

I actually talk a lot of patients out of it. I send a lot of patients away.

(Dr Yard)

I think I dissuade 50 per cent of patients. I use a process of gradualisation ... now is not the time, read the book, think about it and come back next year or in 5 years. Some go away extremely grateful and some go elsewhere.

(Dr Black)

in fact I keep a little book, which I've started in the last year, of people who I've talked out of cosmetic procedures. I talked out quite a few if I think their unsuitable or the risks out-weigh the benefits or, particularly, if they're doing it for the wrong reason.

(Dr Metre)

Refusing to carry out a procedure was not an easy option for doctors, as it took more consulting time and was emotionally difficult for women to accept:

> the hardest thing is to ask a patient to not have surgery. This is the only branch of medicine that I've ever come across where the patient is absolutely upset, horrified that you don't operate.
>
> (Dr Green)

> it is very difficult to refuse when you are in consultation. It is easy to go along with whatever the patient wants, but when you say you won't do it the patient asks why not and they start negotiating with you and it makes it hard. I've got a busy clinic and it takes four times as long in the interview, because often patients see it as a personal rejection.
>
> (Dr Trim)

As mentioned earlier, doctors accepted that most of the women who they refused to operate on would go elsewhere and seek out another doctor to carry out the procedure.

Other doctors saw the consultation as more than just negotiating an aesthetic outcome and carrying out the surgery, there was, in their view, a counselling role. For instance:

> I always give them a very frank opinion because what I'm paid for is threefold. One is to give them a crying shoulder. People who separate, are divorced, bereaved, job problems, economics, everything else – they need a shoulder to cry on, they need to be given hope, if you can give it to them. The second thing is that they need to be given advice like an old-fashioned doctor ... And the third thing, if they want me to execute the surgery.
>
> (Dr Green)

> What comes out of that interview with a patient when you actually take the time to talk to them about their life. What they expect from surgery is that 'stress' comes out and they can be very unhappy, they can cry, they need somebody to listen to them.
>
> (Dr Almond)

This is linked to the psychological aspects of cosmetic surgery which is not just about changing the way the body looks, but also about alleviating unhappiness.

This more holistic approach was recognised as being important by women too:

> I think it would be a good idea that they should have some kind of counselling service before you have surgery and after. It would be good to have that holistic approach in the practice, to make sure people are clear they need and want the surgery because there can be some psychological reasons behind it.
>
> (Chloe)

Taking time and establishing rapport

There was an appreciation by some doctors that spending time and establishing rapport with patients was important:

> The bottom line is you've got to spend time with the patient and let her/him understand and have some sort of a rapport.
>
> (Dr Salt)

> if I don't feel that I have a good rapport with somebody. If I feel that we're just not hitting it off. If I feel that they perhaps don't have confidence in what I'm suggesting, or I have the impression that they don't really understand everything that I'm saying, then I will be reluctant to operate under those circumstances.
>
> (Dr Pound)

Such descriptions suggest that rapport is a two-way process and that doctors, as well as patients, need to feel comfortable before they proceed to surgery. However, Dr Pound's comments also suggests that patients who take the advice of the doctor are more likely to be seen as suitable candidates for surgery. Davis (1995: 117) comments that:

> The medical profession has a long and chequered history of making decisions on women's health 'for their own good'.

Deciding on how a nose or breasts or a facelift should look is another example of such control over women's bodies. Establishing rapport with patients in this context may be more about being able to exercise such control over the aesthetic outcome of the surgery and making the patient comfortable with the vision the doctor has of the changes that will be made to her body.

Conclusion

It is hardly surprising that good communication between doctor and patient is a vital component of successful outcomes in cosmetic surgery. Like other areas of medicine, good doctor-patient communication facilitates a shared understanding of the goals and implications of medical interventions. Unlike other areas of medicine, however, communication in this setting is weighed down with women's experiences of, and feelings about, their bodies, as well as the doctors' reading of those experiences. Doctors are also treading an uncertain path in trying to remodel women's bodies. Successfully negotiating the outcome of surgery is, therefore, fundamentally important to both women and their doctors. While both groups focused on the importance of this negotiation, there were different perspectives about the process through which this was achieved. Patient contact with practice staff, other than the doctor, was significant and was met with a mixed response. Using practice staff as substitutes for tasks that the doctor would usually fulfil, obtaining consent to surgery, answering medical questions and so on, was not acceptable to the women interviewed.

There was ambivalence among both doctors and patients about the role of computer imaging as a communication tool. Some doctors used it regularly while others felt it was not useful. Similarly some patients felt that computer imaging was useful while others saw it as giving an unrealistic view of what was possible through surgery. While women's accounts of computer imagery varied, they were influenced by whether the outcome of their surgery was successful or not.

Apart from the women who had complained about their procedures, women who had received breast implants were the most critical about the outcome of their surgery. All but one of these women had received implants that were larger than they had wanted and all the implant operations had been carried out by male surgeons. Only one of these women had complained about her treatment, and that was because communication had broken down significantly between her and her doctor. For the other women, communication was still good with their doctor, even though they felt that the implants were too large.

It is hardly surprising that communication between the women who complained about the outcome of their surgery and their doctors had broken down. Their stories highlight the fundamental importance of good communication between doctor and patient. They also underline the fact that women who seek out cosmetic surgery want to spend time communicating with their doctors and not with a range of practice staff. They do accept that these staff can substitute for the doctor in some things, but not in the fundamental issues that women felt only the doctor could address.

Women appreciated those doctors who spent time with them and listened to them, and were critical of those with whom they felt rushed and who did not seem to listen. It was clear from what many women reported that the doctor plays a significant role in persuading them or dissuading them about surgical interventions. Of course, part of this role is aimed at achieving a good aesthetic and technical outcome. Women needed to like and trust their doctor and feel comfortable in the consultation.

Doctors had a more structured or task-orientated approach to their communication with women. For women, good interpersonal communication with the doctor was integral across the whole process. While it was also important to doctors, there was more of an emphasis on communication being first and foremost a tool to negotiate the outcome of surgery. Women had a more holistic approach and valued communication in its own right as a mechanism for building trust as well as a means for deliberating about the outcome of surgery.

8
A Risky Business? Understanding Risk in Cosmetic Surgery

This chapter is about how risk is perceived by women and doctors in cosmetic surgery. It will unpick women's narratives and experiences of risk in cosmetic surgery and will explain what women identified as the risks of cosmetic surgery. It will also detail the risks these women reported that they were apprised of by doctors when they decided to have cosmetic surgery and also how these risks were explained to them. Doctors perceptions of risk and how they outline these risks to their patients will be analysed as will the process through which doctors seek informed consent. By analysing what each group says about risk, the chapter will further develop a key theme of this book which is that women and doctors seem to engage in different and separate discourses about the processes and actions involved in cosmetic surgery.

How risk was communicated: Women's perspectives

The ways of communicating risk to women in this study seemed to vary considerably, as did the identification of what constituted a risk. Communicating the risks and obtaining consent was done in a number of ways. Some doctors communicated the risks to patients themselves, some used written materials, some used computer programs, and others transferred this responsibility to clinic staff. The communication of risk was often not systematic, and some doctors were more effective at communicating risk than others. As we saw in Chapter 7, this is not surprising because some doctors are far better at communicating with patients than others seem to be. Patsy, who had consulted more than one doctor, contrasted her experience with two different doctors:

> I needed time to think about this and I needed more information ... (from the first doctor). Both totally different. He was very nice (the second doctor). I felt perfectly safe with him and he asked what I wanted to look like. Even though there was a risk, a very small risk of blindness, I trusted him.

Indeed the inconsistencies in the way risks were outlined to women and how those risks were explained were striking. When women reflected on their lack of adequate understanding of the risks of the procedures they were about to undertake, they often blamed themselves for not knowing what questions to ask. The opportunity to balance risk against the outcome of surgery seemed missing in women's explanations of how risk was discussed. Thus, none of the women seemed to expect that the aesthetic results would not be what they wanted. They seemed to embark on the cosmetic surgery journey with an optimism that was supported by scant information. This information gap was not sufficiently addressed either in the means through which risk was discussed or in the substance of those discussions.

Information about risk given in the consultation

Women's accounts of how information on risk was conveyed to them by their doctors indicate that it was not, in their eyes and at the time of their procedures, a particularly significant part of the consultation. The time spent in the consultation on discussing the risks of the procedure was not significant enough for women to remember it as an important part of the consultation:

> He did explain the risks involved, but he didn't spend much time with us. He is a very busy doctor.

> (Charlotte)

Often women had little memory of what was discussed:

> I don't know. My memory's not so good about that. I don't know. I really can't remember him saying anything about risks. But I had read quite a bit about it and I'd got a book, a paperback, called *Skin Deep* which says there can be risks if you get infections.

> (Irene)

Women reported a range of ways information about risk had been offered to them. They had been given leaflets, typed sheets, made to read information on a computer or delegated to support staff to be told of risks and complete informed consent forms. This variety in ways of outlining risk took the onus away from the doctors to personally explain risk and to obtain informed consent. In general women reported that only a minimal amount of time was spent in the consultation *with the doctor* talking about the risks of the procedure:

> I wouldn't say he explained the risks to me. He put me in front of a power point presentation on the computer and I had to press the enter key after I had read each page. This was the bulk of the second consultation. He left the room and I think went to see other patients. I think his practice is to see a number of different patients at once which seems quite amazing to me.
>
> (Hannah)

Here, the doctor transfers the responsibility of achieving consent to the computer and uses this as the communication tool. Hannah went on to say that

> I made a note of the risks that the presentation outlined and asked the doctor about them when he returned, but he sort of brushed it off, he said it didn't apply.
>
> (Hannah)

Here, the communication of risk by the doctor, or in this case, the computer, has failed to satisfy Hannah and when she seeks more information the doctor is unwilling discuss it.

Written material

Another way risk was identified was through the provision of written information. However, this written material sometimes omitted the particular risks of some of the procedures. In the case of liposuction, some women reported that a significant issue, the wearing of a pressure girdle after liposuction, was omitted from that literature and not outlined verbally to the patient:

> He didn't inform me of the risks verbally. I received some literature, a one page typed-up sheet and a pamphlet. He didn't inform me about the special girdle I had to wear.
>
> (Freya)

I wore the bike shorts for six weeks like he told me to and a couple of months later, I found out about this special garment that was supposed to be worn. And all he said that what I needed was bike shorts. He didn't tell me about special garments or anything.

(Hope)

Whatever information was on the sheet, or in the pamphlet, regarding the risks involved in the procedure, it was not as significant to Freya and Hope as was the information about needing to wear a pressure girdle after the procedure. Providing information in this way, rather than verbally, does not provide patients with the opportunity to seek clarification about the risks from their doctor.

More effective use of written information was noted by Ivana. She was informed of risks by being given a book to read and was then able to formulate questions based on that information:

The risks were outlined in a book given to me by the doctor and when I was in the consultation I then knew the questions to ask.

Being provided with written information before the consultation enabled Ivana to have an informed discussion with her doctor about the risks of the procedure.

Information seeking

Knowing what questions to ask was a recurring theme mentioned by women. A number of women felt, in hindsight, that they had not had enough information about cosmetic surgery to have an informed discussion with their doctor. There were women like Megan who went to some trouble to seek out this information, but it was not forthcoming from her doctor:

Basically, I think I got the information I wanted just because of myself – not necessarily because of him.

There was a view that a family practitioner referring them for surgery might be able to outline some of the risks involved and provide information that might help them seek information from their surgeon:

I'd say (if asked for advice on having plastic surgery) for them to be referred by their Doctor (FP), to ask their Doctor for a referral, and to ask what the risks might be – or the possible side effects. I think

I hadn't really thought about that, I just trusted too much. Mind you, it's all very well to say that with hindsight but how do you think of that beforehand. It's like you don't know the right questions to ask. You don't really know what to ask.

(Molly)

Some women reported that they had tried unsuccessfully to see their doctors to ask questions that may not have been answered in previous consultations. Dora felt that she was 'a nuisance' and that she should just acquiesce to the authority of the doctor:

I felt that I was a nuisance asking those particular questions at that time, but I did explain that I hadn't been able to see her beforehand. Anyway I convinced myself that she knew what she was doing because she said so often enough and I did feel a bit of a goose, so I thought I would try to relax and get through it.

Often in women's narratives it was clear that they had little real knowledge of what to expect from cosmetic surgery. Many women came away from the consultation with the doctor with a number of unanswered questions and relied on their trust in the doctor's skill and knowledge as reassurance that everything would turn out well.

Clinic staff

As mentioned earlier, some doctors transfer the responsibility of outlining risks and obtaining consent to members of the clinic staff. However, this process seemed often to be flawed in that consent was given without full disclosure of risk. When asked whether she had been told about the risks of her surgery by her doctor, Dora said:

No not at the first consultation, that came later. The doctor didn't explain anything. It was one of the girls who said there were risks. But I didn't expect what actually happened to me.

Here Dora was expected to trust the experience of the doctor and also to believe what the clinic staff told her about risks. She had not been informed of possible risks at any stage when consulting with the doctor.

Other women sought clarification from both the doctor and other members of the clinic staff about specific aspects of their procedure, but felt they had been given wrong, or incomplete, information by them.

Some women sought out this additional information about risks to allay their concerns before going ahead with the procedure. The advice they received may have satisfied them before the surgery, however, their experience after surgery seems to vindicate their initial concerns about the surgery:

> No! No! A lot of questions I asked was 'is it painful?' and even after I'd spoken to him and I went out to the nurse, she assured me that 'it doesn't hurt'. I'd seen it done on TV and it didn't look very nice and I said to her 'what's it like. Did you feel anything?' and she said 'Oh no'. So I went in there not expecting to feel anything and obviously I was in shock.
>
> (Hope)

> He kept telling me it was so much worry over nothing. Because the whole day was a bit of a mess. I wasn't really nervous because I had spoken to this receptionist, she was saying to me that it was nothing really … The sedation wasn't strong enough so I was aware of everything they were doing. I could actually feel him sewing my eyes together and they had to give me another dose. But as soon as I looked in the mirror I knew it was wrong.
>
> (Holly)

This minimisation of the risk of pain and discomfort of their procedures and the implication that they were being overly concerned led these women to be angry and upset about their experience of surgery. In contrast to this, another respondent felt that her insistence on seeking out more information on the risks of surgery and on the after effects of the surgery had probably saved her from a surgeon that she had trouble communicating with. The surgeon, having found out that she had sought information from the hospital refused to operate:

> But as it was getting closer to surgery I was getting more and more anxious and that's why I was ringing him and his receptionist was really rude and that's when I rang the hospital … and the nurse said that I should ask my surgeon. I said that he wouldn't answer questions so she put it in my notes. So on the day I was booked into the hospital and I was ready to go into theatre, I was about eight minutes off going into theatre and he came in and had a go at me. He said 'how dare you undermine my decision' and that just shattered me. I had organised time off work, organised the kids, the housework, everything and I thought 'he is not

going to operate'. He kept walking out of the room and he walked back in and said, 'I've made an executive decision. I'm not going to do you today. You can come back and see me another time'.

(Lorna)

This altercation shows a surgeon who had not communicated risk or informed his patient enough to satisfy her. Lorna's attempts to have her questions answered by others seem to have been regarded by him as a challenge to his authority and he, therefore, used his more powerful position in the relationship and refused to operate.

What is evident is that women often went into surgery without being told of all the risks involved in the procedure they were about to undergo. What is also evident is that doctors looked for ways to outline risk to patients that did not involve spending time in consultation with them. While not all patients were dealt with in this way, a significant number pointed to real problems in the methods used to outline risk. The means through which the risk message is given may add to the perception, widely touted in the media, that it is not 'real' surgery. Those that have analysed such publications found that the benefits of surgery were emphasised and the risks minimised (Sullivan, 2001). Women go into the process of cosmetic surgery often not understanding risks they might face. This perception is supported by doctors who fail to comprehensively explain risks and whose methods of explaining risk are often at odds with what is understood as traditional medical practice.

Of particular note here is that women who had complained about the results of their surgery were more likely to say that the risks of the procedure had not been outlined adequately to them. Nevertheless many of the other women interviewed also noted that certain risks were only sketchily outlined to them, if at all:

he didn't tell me about the possible negative consequences. It might have affected my decision. I'm really not sure. It might have put me off and I guess that's why he doesn't say the negative consequences because he doesn't want to put people off.

(Molly)

It was a bit of a shock when I saw myself after the surgery because I was black and blue. I don't think he quite went in to that. I don't think he wanted me to know how bad I was going to look.

(Irene)

Molly and Irene were ambivalent about whether they would have gone ahead with the procedures if they knew about the 'negative consequences'. They attribute this to doctors not wanting to 'put them off'. It suggests a paternalistic attitude among the doctors who fail to provide information to adult women who are capable of making their own decisions about their bodies. The choices women make about their bodies, under these circumstances, are made without a fully informed understanding of what the potential consequences may be.

While some felt that their doctors kept some information from them so as not to 'put them off' surgery, some were clear that they were so keen to go ahead with surgery that information about risks was unlikely to dissuade them:

> Risks of anaesthetic, but nothing over the top. I was very keen to have it done. He said I should go away and make up my mind, but I'd already made up my mind. I just walked out and made the appointment at the desk. ... I don't think anything was going to stop me doing it, so whatever he said wouldn't have mattered, I was determined to do it anyway.
>
> (Grace)

Few women remembered in detail all the risks outlined to them by their doctors. It was more likely that, as the following excerpt indicates, they remembered certain risks which they regarded as significant:

> Yes, talked about heaps of risks. He talked about different infections and dropping of implants, that's all I can remember but I'm sure he went into other ones but those were the important ones to me.
>
> (Thelma)

Overall, it was the time the doctors spent talking about risks and the importance they placed on outlining those risks that were of significance to women. Some written materials were useful, but others were not detailed enough and did not address significant and specific risks. The use of practice staff to detail the risks to women was not a successful strategy. Either staff could not answer questions, or they played down the significance of risks by reassuring women of the skill of the doctor, implying that such a negative event was not possible with a skilled practitioner.

What risks were outlined to women?

Women reported that certain risks were outlined to them, but not others. In general, the risks outlined seemed to be those associated with an adverse event that can occur in any surgical intervention. Risks associated with particular procedures, problems that may arise when recovering from a procedure, or the risk that the aesthetic outcome may be unsatisfactory to the patient, were usually not dealt with. For instance Patsy talked of certain risks being outlined but not others:

> He told me only of the risks of anaesthesia, he didn't talk about the risk of the implants hardening, and he didn't tell me I should massage the breasts.

Significantly here, it is a well-known and real risk that breast implants can harden or encapsulate, but this information was not conveyed to Patsy.

Also of concern was post-operative recovery and many women identified this as a risk that had not been outlined to them. Donna had contacted a number of doctors before opting for liposuction and compared their approaches. All but the one she finally used said she could go home later in the day after she recovered from the anaesthetic. The doctor she used suggested she stay in hospital overnight. She said:

> It was the best thing I ever did ... I was like a sieve ... I was just oozing watered down blood. I thought there is no way I would have coped if I'd gone home, because it just oozes through the bandages.

However, Freya who had undergone the same procedure was not as well informed:

> This is how bad the communication and information was, he just told me that I would need to wear something tight after the surgery, he didn't explain that I had to wear a specific girdle, so I took along tights and there I was on the day and I didn't have the right garment. So I had to go out the day after the surgery and buy one and I was all bruised and swollen and I had to put it on, there was nothing in the information that specified what I had to wear. Also, he didn't mention about the oozing after the surgery beforehand. After the surgery, a nurse said I would ooze but I was still coming out of the anaesthetic.
> (Freya)

Such side effects had caused significant discomfort and distress to these women but were dealt with very indifferently by their doctors.

Many women commented that they felt they had been left in the dark about the post-operative phase of their surgery and were not expecting the discomfort and pain that they experienced. Davis (1995: 148) reports similar findings from her research:

> In retrospect, several women complained that they had been poorly informed about the discomfort and side effects which often accompany breast implant surgery. They did not know until several weeks after the operation, for example, that massaging the breasts could help prevent the implants from becoming hard. Others explained that while they had heard about the risk of encapsulation, they had no idea what to expect. They were completely unprepared for the spectacle of a hardened implant. ('if I'd seen a picture, I might have had second thoughts.') Nor did they know what to expect when they came in to have the encapsulated implants palpitated. ('I thought I was going to die, it was so painful.')

This theme was echoed by many women in this study:

> He didn't explain any of the risks about encapsulation to me or infection but when I went to see the other surgeon who did the second operation to fix me up he gave me all these brochures and all the risks were there in black and white.
>
> (Patsy)

> I would have a thousand children rather than go through that again. It was very painful. I didn't expect it to be that painful. It was very very painful for the first couple of weeks and then painful for a couple of months. Even now, 11 months later, one of them is still very uncomfortable.
>
> (Thelma)

Fiona also had breast implants but she was told about the risk of encapsulation and told what she needed to do to help stop it. However, she describes the pain experienced when massaging the breast to reduce the risk of encapsulation:

> He told me exactly what he was going to do, do an incision under the armpit and insert the implant into the pocket behind the breast tissue. I would have to move the implant around three times a day for three months after that and then once a day for six months to stop the implant getting hard. I had to get my husband to do it for

me because it was so painful. I honestly don't know how anyone could do it themselves.

(Fiona)

The example of the risk of encapsulation of implants offers a good illustration of how some doctors did not inform their patients of a significant and proven risk associated with a particular procedure. Given the importance of manipulating the implants in order to avoid encapsulation, it is surprising that this risk was not emphasised by every doctor that carried out the procedure. One explanation for this inconsistency is the pain that results from massaging the breast. This is not something that fits in with the image projected by many practitioners of cosmetic surgery as a seamless transition to a more beautiful body.

Women also talked of how they felt in the days after their surgery. They were not expecting the pain, discomfort and swelling that they experienced. Some talked about their doctors saying that they would be 'back at work in a couple of weeks' after surgery. Many took weeks to recover physically and were unprepared for the emotional effects of the surgery. They may have been satisfied with the outcome, but felt that they had not been well informed about the post-operative phase and the cost to their body and emotional consequences of what they had gone through.

The doctors' reticence in outlining the likely physical and emotional consequences of surgery could be because of a range of factors. Firstly, it may not occur to them that the post-operative phase can be so difficult. Also, it may not be in their interests to emphasise too much the negative consequences of surgery, beyond what is strictly necessary, as patients may not be as keen to proceed. At the same time, women like Molly who were keen to undergo a procedure may be less receptive to talk of pain and discomfort:

He did explain what was going to happen and, in fact, part of me wished that he wouldn't because it sounded yucky and I didn't want to know about it. But I think, as I said, he didn't tell me about the possible negative consequences. It might have affected my decision. I'm really not sure. It might have put me off and I guess that's why he doesn't say the negative consequences because he doesn't want to put people off.

(Molly)

As will be discussed later, the doctors interviewed, while being adamant that they did outline risks to patients, saw it as the risk of a poor outcome due to 'unrealistic expectations' or the risk of an adverse event, through anaesthetic, bleeding etc. Women, on the other hand saw risk as including other post-operative effects. Some had sought out information but had been reassured, even though their post-operative experience was at odds with these reassurances:

> he didn't explain to me how bruised and horrible I'd look afterwards. He made a bit light of it. It was a bit of a shock when I saw myself after the surgery because I was black and blue. I don't think he quite went in to that. I don't think he wanted me to know how bad I was going to look.
>
> (Irene)

Others found out about possible serious side effects after they had undergone the surgery. If they had been told about these risks before the surgery, they may not have undergone the procedure:

> No he didn't (talk about risk) and I was a bit cross about that, because when he came to see me after I had come out of the anaesthetic he said 'Oh it's really good. There's been no nerve damage'. And I thought 'nerve damage?! He didn't tell me before that there might have been nerve damage!' So I was cross about that. I didn't say anything because I thought it's too late now and he had told me I hadn't got any. But I do think that I went in to it a bit naively and I do think if I had to do it again I wouldn't go in to it that naively.
>
> (Molly)

Overall, the evidence presented by women in this study shows that many practitioners do not systematically and thoroughly outline risks to their patients in a manner or form that patients easily understand. When doctors are systematic in outlining risk, and offer patients enough information for them to make meaningful choices, their patients appreciate this and are prepared to be forgiving of outcomes that are not what they may have expected. In general, people are cognisant of the major risks associated with surgical procedures. Where cosmetic procedures are concerned, these risks are no less real. However, the risks may not seem as real due to the discourse that surrounds cosmetic surgery, a discourse that underplays risks in order to promote the industry. Thus, the media and advertising often frame cosmetic procedures in a way that belies their medical basis. It portrays cosmetic procedures as a lifestyle choice rather than a medical intervention. Women who access the procedures often are, or feel that they are, regarded as vain and frivolous. Many who undergo cosmetic procedures do so in secret and are reluctant to tell anyone but their closest family or friends. Because of this, it is important that they slip into and out of the procedure in as little time and with as little fuss as possible. In many other medical procedures they undergo, women enter into the process relatively well informed. There is information relating to their medical needs out in the public arena and available through the networks they have. Thus information on women's health issues is shared by, and among, women. For cosmetic surgery patients, this network is either not there, or only comes to the fore after they have undergone their procedure and have 'come out' to others. There is no independent pool of knowledge available to women that may inform them of potential risks.

The risk of an unsatisfactory aesthetic outcome

Women's individual situations and motivations also have an impact on the way they approach the surgery and assess risk. Thus:

> While there may be an awareness of potential risks the flip-side of taking those risks may be a feeling of alienation or deficiency to other women. With that comes the risk of unhappiness, lack of self-esteem and the inability to fully engage with living. Sometimes the social risks were more certain because they had already been experienced. Indeed, even the negative moral judgements commonly associated with cosmetic surgery were not enough to outweigh the suffering from appearance concerns.
>
> (Rowsell et al., 2000: 210)

In undergoing cosmetic surgery, women are reacting to the risk they face every day of the consequences of negative feelings they have about parts of their body. This known and understood risk contrasts with the unknown and poorly understood risks of cosmetic surgery. When women choose to have cosmetic surgery, they expect that the risk of undergoing a procedure is less than the risk of living with their bodies as it currently is. However, accepting the risk of an operation or of postoperative recovery is very different from accepting that the result of the surgery might not be what was wanted or expected.

It is clear that one of the most significant risks that women face when undergoing a cosmetic procedure is that the aesthetic outcome will not be what they want or that it makes them feel worse about themselves than it did earlier. Those women who had complained about the outcome of their surgery felt this most keenly. The extent of their distress with their 'new' bodies was palpable, as was their feeling that the ability to control how their body looked had been taken from them. While some of these excerpts have been provided in earlier chapters, it is useful to return to them here. Patsy, Lin and Dora clearly describe the way they felt after their surgery:

> It was like he was turning me into the vision of his perfect woman. He didn't listen to me, he wanted me to look a certain way and that's what he did. He didn't do what I wanted. He did what he wanted ... It changed my life totally. It started when I woke up and I had these enormous breasts and the surgeon just said to me " I made them a bit larger, because you'd only want them larger a year down the track". They were a double D and I was supposed to be a C cup. They were just vile. I ended up having more surgery in June last year to remove them and have smaller implants.
>
> (Patsy)

> When I look back I wonder why I put myself through that. Now there was definite asymmetry between one eye and the other. I couldn't stand myself.
>
> (Lin)

> Afterwards she said that I looked much younger and I told her that I hadn't come there to look younger I just wanted to look tidier, but by then we were at cross purposes.
>
> (Dora)

These women now had bodies that they were even more at odds with. Added to this, they felt angry and disempowered by the way their wishes had been misinterpreted or ignored by their doctor. These women's right to make decisions about their bodies had been taken from them by their doctor and their wishes either ignored or reinterpreted.

However, there were women who had not complained about the outcome of their surgery who felt that the result of their surgery was not quite what they had expected. As Ivana said:

> I felt sort of cheated in a way ... It's not just financially, it's emotionally. If it's not the result you after, it's sort of like being stuck with a bad haircut, but it's not the bad haircut that can grow out or change.
>
> (Ivana)

It was the emotional cost of an unwanted outcome that was felt most keenly by these women. Not only did they feel bad about their body and angry about their treatment, but they felt ashamed of being 'cheated'. As mentioned in Chapter 5, some felt that there was very little sympathy for them when they approached other doctors to have their problem remedied. Dora did report having a more sympathetic response but nothing could be done to remedy her problem:

> This doctor said it couldn't be altered so I just decided to get back on track and get back to work.
>
> (Dora)

Even though women were angry at the doctors who had treated them, some felt that they were being punished for their vanity:

> It brings home to me that you don't play around with what you've got and I was given a pretty fair deal to start with and now it's like I've been punished for mucking around.
>
> (Lin)

Self-blame wove its way through these women's stories. They felt that they should have been more alert to potential problems and less trusting of their doctor.

The particular case of breast implants

A significant issue that was mentioned by women who had breast implants was that their implants were larger than they expected:

> They are bigger than I expected them to be and I did mention that at my first check up with him, but he's not interested in hearing that. As far as he was concerned I looked fantastic and there was no infection. I'm a D but I would have preferred to be a C.
>
> (Thelma)

The way in which women and their doctors negotiated the size of implant was discussed in Chapter 7. Even though women reported a range of ways in which this negotiation took place, all but one of the women who had breast implants said that those implants were larger than they wanted. Before surgery these women felt they stood out because they were flat-chested. After surgery they stood out because their breasts were too large and were now 'like headlights' (Thelma) and attracted unwanted attention.

Another woman had her implants removed because:

> People used to look at me differently. They'd just look at my boobs and not at me.
>
> (Patsy)

These stories of large implants suggest that some doctors objectify women's bodies and insert implants that are more in keeping with their notion of the ideal female form, rather than their patients'.

Even though there has been much publicity about breast implants, many women interviewed were not well informed about, or did not appreciate fully, the post-operative consequences of having the implants. These consequences included the pain and discomfort associated with the implants, the feeling of heaviness of the implants, and the risk of encapsulation. Interestingly, only one of the women discussed the litigation involving silicone implants. The implants that she had originally had were silicone. These were now 'under the bed' pending the results of the class action against the implant manufacturers:

> they started going hard and misshapen, there was a number of operations but there was no way they were going to get any better. Anyway my doctor moved to Israel so I just put up with them. I couldn't lie face down. It was too painful to lie on them. I decided to have them removed within about 3 years but didn't get around to it because there was a cost. When I decided to have them out, I went to my GP and said I didn't want a rude, arrogant plastic surgeon, which seemed to cross 90 % of them off the list. In the end I ended up with a guy up the road, who was gorgeous. I said that I wanted them out; he said he'd removed heaps. He said he would do it at a reduced cost because he felt he owed something back for all those women who had it done and didn't get what they wanted.
>
> (Stella)

The story Stella tells is of suffering discomfort and pain from her implants and of being rescued by a sympathetic doctor who was used to seeing women whose implants were not what they wanted.

Patsy and Sonia also felt their implants to have been painful and to be 'different' and alien. This foreign feeling about the implants was clearly something they had not expected and they talked about them as if they had somehow taken up unwelcome residence in their body:

> No it was awful. I had a lot of pain then and he said don't worry about them. In fact I went back a year later and said 'Take them out, I hate them'. I was in tears.
>
> (Patsy)

> but the way the implants feel is different. As far as the implants go ... they feel heavy and they feel different.
>
> (Sonia)

Lorna was the only woman who had wanted larger implants but did not get them. She explained how she had negotiated the size of implant with each surgeon:

> Yes (I wanted) big and obscene. He talked me out of that. He didn't sway my ideas; he just said that the implants I had originally picked were very big. So I said on a medical term or on a personal term and he didn't really answer that. He said that they would be very big for my frame. The first surgeon didn't discuss that with me. He wouldn't discuss it with me and the implants I ordered from him were 470 ml with a tear drop shape. This surgeon asked why I'd ordered a tear drop and I said to fill out the sag. I said I'd been quoted an extra $4200 and he was very surprised. He said that he wouldn't charge me anywhere near that. He would charge me an extra $500. He said he wouldn't cut away any excess sag. He explained that once the muscle was lifted back up I would get a lot of the fullness back so there would be no need to go for a really big implant because I would look really top-heavy. I was then going to go for the 400 ml and then I tried a 380 ml on, because I'm sized seven from the waist up and it would have looked too obscene and even though now I would have liked them to be a bit bigger, but it does look natural.
>
> (Lorna)

Although Lorna had smaller implants than the 'big and obscene' ones she requested, it is still the doctor who decided about the size of the implants. While this opinion may be justified on medical grounds, the doctor did not offer any medical reasons why this woman could not have larger implants. His reasons, 'they would be big for my frame', is a comment based on aesthetic and not medical grounds.

Additional procedures

Another area of concern for women was that they put themselves at risk of embarking on more surgery or more extensive surgery than they actually felt they needed or wanted. Women did not realise that this was a possibility before surgery; but for many it became a reality either when they consulted with a doctor or when they woke up after their surgery.

Some of the women who sought facial surgery reported that they had been offered additional procedures to the one that they were originally seeking. For instance, it was suggested to Lin that she should have blepharoscopy (eye lid surgery) when she went in to have laser treatment for the areas under her eyes:

> my family suffers from a hereditary condition where we have dark freckles under our eyes. Anyway I was lined up with a top photographer and I could hear them say to the makeup people 'look do up the freckles under the eyes'. After that I developed a complex about it and I read about a laser procedure in a magazine that could eradicate dark circles under the eyes. I got a referral ... I went to see her and it was a very brief consultation, I explained about the pigmentation. The consultation was 5–10 minutes maximum. She sat me on a bed and had a look and she also suggested I undergo a blepharoscopy ... I didn't feel I had a problem and I questioned that and she said for optimum results it would be best if we did that also.
>
> (Lin)

Given that Lin was only in her twenties, it is difficult to understand why such surgery would be necessary and how it would improve the dark circles under her eyes.

Chloe had a similar experience when she went back to her doctor because the liposuction that had been carried out on her face was uneven. The doctor suggested that the real problem was to do with her eyes:

> 'Look don't worry about that. You need your eyes done. Your lids are hanging over your eyes like curtains. Come and get your eyes done'. And I thought, 'Oh my god now I need my eyes done as well, and I thought, that's not what I want to do, I want to fix this' (pointing to the face).
>
> (Chloe)

Again the effect of this was to make her feel even more dissatisfied with the way she looked.

Charlotte and her friend had decided to have rhinoplasty together. In one of the consultations the doctor suggested that her friend also have a chin implant:

> One of them offered my friend a chin implant.
>
> (Charlottte)

Her friend did not have the money to do this and was not '100 per cent happy' with her surgery. The implant may have improved the aesthetic outcome of the surgery, but the suggestion by the doctor that it was needed may also have made this young woman feel further dissatisfaction with the way she looked.

More significantly, women like Dora were offered additional procedures that they felt they did not need and that were more invasive than they wanted:

> Well she said I needed the face lift, the resurfacing and the brow lift. I only wanted a bit of a tidy up and I didn't go ahead with the resurfacing.
>
> (Dora)

The story Dora tells is of control over her body being taken away from her. She had wanted laser surgery but was convinced into having a facelift. However, she had not realised the degree of surgery that this entailed and only began to appreciate what was going on when she was about to be operated on:

> When I was lying there on the table I had given up arguing or discussing it with her and she started marking out and she was drawing all these lines and I thought, 'my god, I'm going under now and there's no more time left' and then she drew this big line around the back and all I could think of was 'what is that for?', but it was too late I was going under.
>
> (Dora)

Molly was offered alternative procedures that seemed to be a sensible option for her:

> I went asking to get my baggy neck not so baggy and the surgeon explained to me that he couldn't really do what I wanted. What he

had to do was what he called a mini face-lift. He said that if he just did the neck, there was no way of hiding the scars. That seemed to make sense to me.

(Molly)

When doctors explained clearly to women what was possible and what was not, women were able to make an informed choice. When doctors did not explain their rationale for alternative or additional surgery, women were left more dissatisfied with their bodies and their attention was now focused on yet another fault.

Such experiences bring into sharp focus the balancing of the risks of cosmetic surgery with the benefits. While the benefits might be that women actually get what they want from surgery, the risk is that they become further alienated from their body by an outcome they did not want, a procedure that goes wrong or the suggestion by their doctor that there is something else amiss with their body that could do with correcting.

Balancing risk and benefit

For women, the risks associated with cosmetic procedures have additional dimensions to those usually associated with other medical procedures. Risk, in this context, is a far more complex and dynamic process. It is linked to how they feel about their body and the possible benefits, to them, of the procedures they want to undergo. The risk of doing nothing has to be balanced against the risk of undergoing a surgical procedure. This, of course, does not mean that they would go ahead with a cosmetic procedure irrespective of the risks involved. Most participants in this research reported that, in retrospect, their conceptualisation of risk was not the same as that of the doctors. They had a much broader view of risk, it included the outcome of the surgery, significant medical or surgical risks that may result in an adverse event and the way they felt both physically and psychologically after the procedure.

When reflecting on what risks they had identified as being involved in cosmetic procedures, many women reported a broader range of risks than did the doctors. Thus, women understood that there were medical and surgical risks, as in all such procedures. However, they also talked about risks associated with the outcome of surgery, including aesthetic results and their physical and psychological health after the procedure.

What these women often painted was a picture of being at 'cross purposes' with their doctors. Many felt that their doctors did not really understand what they wanted to know about the procedure, and how they wanted to look after the procedure. They also felt that they did not know what questions they needed to ask to fully inform themselves about the procedures they were to undergo:

> I feel that there was a failure to provide information, the fact that I would have these effects was simply negated from the conversation. I suppose, looking back, I just didn't know what questions to ask.
>
> (Hannah)

The risk of an adverse event was well understood by women in general terms. They saw them as the risks that are associated with any surgical procedures. These risks seemed to be confined to what happens at the time of the procedure due to anaesthesia, bleeding etc. rather than what can happen afterwards. The Australian legal case of *Rogers* v. *Whittaker* (1992) was legal authority for the principle that a material risk arising out of a procedure that was real, even if it was statistically highly unlikely to occur, should be outlined to a patient. In the interviews with women, there was no evidence that doctors explained such statistically small risks, nor indeed, did they outline the risks most likely to result from particular procedures. For the women who had experienced a poor outcome from their surgery, this was particularly significant.

When a procedure did have a complication that could have been foreseen because it was a risk in that type of surgery, women felt that they should have been alerted to it prior to the procedure being undertaken. The information might have made them more cautious about having the surgery and less angry with the doctor when things did go wrong. Like other patients in the health system, women who access cosmetic surgery wanted comprehensive information about the procedure they were about to undergo and all the risks that were attached to that procedure. If things did go wrong, they were less likely to be angry at the doctor and less likely, therefore, to seek retribution through making a formal complaint.

How do doctors explain risk and obtain informed consent?

> I had my own cosmetic procedure, Rhinoplasty, done many years ago, I was so badly informed and so badly looked after, I had a result which was to me not good so I thought there has to be a better way.

That gave me the incentive, because I believe that women undergoing cosmetic procedures, should be not only better informed but the
standard also needed to be a lot higher.

(Dr Metre)

This statement from Dr Metre acknowledges that the provision of
information to patients may be inadequate and incomplete. However,
this view was not shared by many of the other doctors interviewed,
although as has already been discussed, it was the view expressed by
many of the women.

All the doctors interviewed reported that they outlined risks to
patients. Some used written material:

I outline risks to them firstly by giving them a book I wrote. I send
them away with the book and then see them again. I see every
patient twice before I operate.

(Dr Black)

they get a pamphlet which outlines, you know, the normal course
and what other things would happen. ... it is harder to get patients
to actually listen to anything.

(Dr Dance)

Although Dr Dance felt that patients did not listen, he went on to say:

I always start off with a question. I always say to them here, 'Now
what could go wrong with the procedure?' I make it a point to question them, I make them think about what can go wrong with the
procedure. And then they say, 'Well like what?' I say, 'Well number
one – and I think I say it with all procedures, first of all bleeding. In
any operation there could be bleeding which could cause bruising
or haematoma or ... and then I add on what we do to prevent that
complication. And we do a lot of things to prevent it. We use drain-
age due to pressure dressing. If the blood pressure is slightly lower,
we do numerous things to minimise bleeding during complicated
surgery. What else do we talk about? Infection is possible with any
operation. So what do we do to prevent it? We always give intrave-
nous antibiotics during the operation and after the operation and
we wash out any catheter with betadine solution. The scarring – you
know where the scars will be. It is very important for them to know
and understand.

Even though Dr Dance reports detailing potential risks to patients, he seems to balance it with explaining how this risk is addressed. This approach may well reassure patients that they are in good hands, but it may also encourage them to believe that all risks can be easily dealt with when they arise.

Similar to Dr Dance, the following doctors said that they explained risks to patients themselves:

> We have consent forms for everything we do in the practice but I make it a point with larger procedures in particular, to outline risks from the first interview and make sure they understand risks.
>
> (Dr Pepper)

> At the end of the main consultation when we are discussing the outcomes, I also discuss the complications that can occur. And I usually do a very detailed discussion on that.
>
> (Dr Almond)

The methods doctors used when trying to explain risk varied. For instance, Dr Step noted that women did not have to remember what he was saying:

> So the first time I have a bit of speech in what I tell people. When they come in I say, 'Look just relax. You don't have to make any decisions. In fact you won't be allowed to make decisions. All we do is talk to you, at the end of which you will have as full idea of the technical side of the recovery phase, risks or if there is complications and so on. I don't discuss finances. My nurse does that and she's excellent. What I say to people, 'You don't have to remember what I'm talking about because we're going to go through the whole thing again'. So I'll go through everything and document it.
>
> (Dr Step)

This may have the effect of making the information he is conveying less significant than it really is.

Dr Cherry commented that risks of smaller procedures weren't usually described well to patients:

> With the smaller procedures, we do tend to, I think we underplay the risks to some extent, because in our own minds they're comfortable

with the procedure not producing a risk. Now that's really quite dangerous, but it is true.

(Dr Cherry)

However small the procedure is, any risk that could be involved should be outlined to women if they are to make an informed choice about whether to go ahead or not.

The issue of whether informed consent was achieved for cosmetic procedures was one that elicited different responses from some doctors. Dr Inch questioned whether, in fact, fully informed consent is ever possible if women do not really understand what they are consenting to:

Whose body is it? Who's in charge of that body and I think that the law clearly states that the patient is in charge of their body but our duty is to get fully informed consent and if you do that you are doing the right thing. Can you get fully informed consent, can they understand what it all means?

(Dr Inch)

Dr Apple noted that obtaining consent was a relatively new phenomenon in Australia:

When I first started practice you hardly did it at all. When you used to go to America and they used to talk about informed consent and we used to laugh and say 'we don't have that'. But it's built up and built up and now we have full informed consent and you try to think of everything that could possibly go wrong.

(Dr Apple)

Although Dr Apple now believes that there is 'full informed consent', he went on to categorise informed consent in the following manner. He underlines the role advertising has in inflating women's expectations about what can be achieved through surgery:

A lot of informed consent is usually reducing their expectations because advertising has usually raised their expectations.

(Dr Apple)

The use of practice staff

Clinic staff were used to provide additional information and to complete the relevant forms. This transfers the responsibility from the doctor to other clinic staff to outline risks and obtain consent for the procedure:

> I speak with patients but then I send them to one of my cosmetic nurses who has a form and asks the patient to sign each point as they've gone through it. So we go point by point by point – in great detail. The nurses will spend at least an hour with the patient going through all the side effects.
>
> (Dr Apple)

> I have a nurse who has been trained in cosmetic consultation, cosmetic surgery and she will go through the operation and its potential complications in detail and then when I see the patient, I will go through anything that is still unclear with them. We also give them information which I have written about the potential complications and I always include disastrous complications, including death.
>
> (Dr Crisp)

Thus, doctors reported that they outline what they see as the risks involved in cosmetic procedures and they do this in a number of ways. Some talk to women themselves, while some hand over the responsibility of filling consent forms and going through the details of risk, to staff. These staff may or may not be medically trained.

What do doctors see as risks in cosmetic surgery?

Doctors interviewed saw risk as falling into three areas. Firstly, the risk of an adverse event which can occur in any surgical or medical intervention; secondly, the risk that the women had unrealistic expectations as to the outcome of the procedure; thirdly, the motivations behind women seeking cosmetic procedures.

Risks involved in surgery

Doctors noted that they discussed the risks that are part of any surgical procedure with their patients:

> First of all there's the medical risk which we spend a lot of time on. As you would have seen by now we have a number of leaflets from the Australian Society of Plastic Surgeons, leaflets which outline all those risks, not always to my satisfaction. I go through that with them as well. When you go through the medical risks some say, 'gee

I didn't know you could die from an anaesthetic' or 'I didn't know that there would be scars' and they go away and never come back. A lot of them go to another surgeon, who doesn't tell them those things, and that's fine. That's between them and that surgeon.

(Dr Inch)

I find it quite important to let them know that that has got as much risk as any other operation because it is an operation. And then usually I explain to them on general complications and anaesthetic complications. I firmly believe cosmetic surgery should not impair one's general health.

(Dr Salt)

Doctors did not say that they outlined the risk of a poor aesthetic outcome to their patients, neither did they talk about the post-operative risks that women might face. The doctor's risk discourse was limited to a scientific and medical paradigm, an objective assessment rather than one that was more pertinent to a woman's subjective embodied experience. That is, given the way women said they felt about their bodies before cosmetic surgery, a minimal risk of a medical disaster may not have been given much weight when compared with the risk of continuing as they were or of getting a very poor aesthetic outcome.

Unrealistic expectations or unsatisfactory outcomes?

Many doctors mentioned that one of the significant risks from their perspective was women having 'unrealistic expectations' as to the outcome of the surgery:

I think the best thing that I can do for a patient is to establish a relationship and I don't promise. I mean I guess there aren't applied promises but I go to enormous lengths to say, 'This is what I can do. This is what I can't do'. And, in addition to all the complications I go through the physical complications. I go through anaesthesia, infections, specific – you know, everything there is. And the final one I emphasize is, I say, 'The biggest complication is unrealistic expectations', and I have a bit of a 'throw out'. I say, 'This will not get you a better husband, boyfriend, girlfriend, job, car. It might not even make you happy. But, if you believe that this is your personal thing (whatever it is), and you want to have something, then I can do that – no more'. And I put it like that.

(Dr Step)

As this quote indicates, the 'complication' of 'unrealistic expectation' is regarded by this doctor as more significant than a possible technical mishap. Women's expectations as to what can be done through cosmetic surgery, as mentioned earlier, was seen to be driven partly by the influences of advertising.

Doctors felt that negotiating what can be achieved is an important part of the consultation. For instance, Dr Green described how he discussed what he felt he could and could not achieve in a procedure in the following way:

> Well I use two things. One is to be realistic and say, 'Well you can't get a hundred per cent result. I can't turn a six foot lady into a five foot three, genteel person. Or turn Lassie into Marilyn Monroe. You know, they're the sort of silly things that you say – but they stick. You say, 'so if you're prepared to accept these limitations on the result, the surgery has these physical problems'. So they are the areas. One is, what can I achieve and I always quote seventy per cent improvement: and the second thing is, are you prepared to wear the following risks – anything up to death in fact. It never happened to me but you know, sooner or later I will run out of luck. 'If you are prepared to pay that risk for that improvement – we've got a deal'.
>
> (Dr Green)

Again the emphasis is on the aesthetic rather than technical issues in the procedure, with the serious complication of 'death' being seen as a gamble entered into by doctor and patient. The analogies used by Dr Green seem inappropriate and flippant in identifying the limits of what he could do. He is offering himself up to the patient as less than perfect and passes on to the patient the responsibility of deciding whether or not to undergo the procedure, even though they may not get the result they hoped for. The 'choice' he offers patients is a gamble, they can withdraw from the process entirely, go elsewhere or accept the doctor's terms and continue with the procedure. However, implicit in this choice is uncertainty. Women are told what he cannot do, but do they fully understand what he can do?

When a doctor approaches the issue of risk and consent in this way, he may be attempting to encourage the patient to be 'realistic' about the outcome, but he is also introducing further uncertainty into the process. The patient will have some image of what she wants to look like at the end of the surgery, but now she is told that this may be 'unrealistic' and that the doctor will only be able to achieve what he/she believes is 'realistic'. Because both the image held by the patient and that held by the doctor are difficult to share, both being in the mind of each individual, the concept of what is realistic and unrealistic becomes meaningless unless it is measured against a concrete yardstick. In the second quote, the doctor noted flippantly, that he couldn't turn 'Lassie into Marilyn Monroe'. Women in this research and in previous research (Davis, 1995) in general, did not want to look like any one else. They wanted to be a better version of themselves. They didn't want to be Marilyn Munroe. The notion of a shared understanding of outcome is discussed by Davis (1995: 129–30) in this way:

> While the women expressed the desire to receive enough information to assess whether the results were going to be worth the trouble, the surgeons seemed more concerned with medical criteria, like function or discomfort. Aesthetic criteria – that is, how the breasts were actually going to look – were of secondary importance. However, even when they did deal with the issue of appearance, surgeons seemed to speak a different language than their patients.

While the doctors interviewed for this study were more concerned with aesthetic outcome than those in Davis' research, the issue of patient and doctor understanding each other is common. It is not surprising that surgeons 'speak a different language'. They are influenced, not only by technical and medical issues but also by their normative standards of what constitutes female beauty or an 'acceptable' female form. In cosmetic surgery a healthy body is defined, by both patient and doctor, as deficient or pathological. What constitutes an acceptable alternative, however, is what is in question. Dr Black commented that:

> John Houston said we should never forget that plastic surgery is the surgery of sex in all its ramifications. It's a very complicated thing. It's all about attractiveness. It seems that physical attractiveness is more important to men than women and the pressure gets pushed towards women. So Nefertiti is a stylised version of female beauty, long jaw line, elegant neck, and the nose I've been creating for years is her nose.

This quote indicates that, whatever a patient might want, doctors will create what seems right to them, or what they are comfortable creating, be it Nefertiti's nose or another 'stylised version of female beauty'. Thus, the changes these doctors make to women's bodies may well be based on subjective judgements that are made legitimate through the use of medical discourse around what is surgically possible and what is not. The doctor's views and actions may also be based on concern for self-preservation or on the doctor's own view of what is in the patient's best interest. Outlining risk to patients to include aesthetic outcomes provides insurance should that outcome be well outside the patient's expectations. It also transfers the onus of responsibility to the patient and may lower their expectations of what can be achieved. There is little visual sharing of knowledge of what the patient wants and what the doctor can provide. Of course, not all doctors act in this way. As mentioned in Chapter 7, many turn to computer imaging and photographs to achieve a shared acceptance of what aesthetic outcome can be achieved:

> usually I talk to them about what the procedure involves, what might be suitable for them etc. Then I talk about the risks that are going to be relevant to them. Then there would be more detailed computer imaging. The risks would be outlined in even more detail if they decide to book into the procedure and they have a pre-operative appointment with M (a staff member) prior to that and virtually every risk is pointed out to them.
>
> (Dr Metre)

However, not all doctors were convinced that computer imaging was superior to pen and paper:

> I usually give them the information sheet that I buy from my society (I usually refer to the society), and also I sit down, draw diagrams and explain to them. Patients ask for other photographs and also other patients photographs. Every patient gets photographs taken ... I don't do that (computer imaging) because you know, in my mind you can do what you like with a computer ... I think you can just draw things.
>
> (Dr Salt)

Dr Inch reframed the issue of satisfactory outcomes in this way:

> People talk about good results and bad results from surgery and I don't think that's really relevant, I think you have happy patients and unhappy patients. Some of my best technical results have been patients who have been unhappy.
>
> (Dr Inch)

Here, he regards a good 'technical' outcome as the measure of a successful result. Patients are labelled as being, at heart, happy or unhappy people rather than being dissatisfied with the aesthetic outcome. Comments like these indicate that an understanding of what patients believe is a good outcome for them can be at odds with the surgeon's view that a good technical result is the measure of success. An outcome may well be technically acceptable but not aesthetically acceptable to the patient.

This distinction is important when the aim of the surgery for the patient is to have an aesthetic improvement. As we have already seen, women who access cosmetic surgery feel that a particular part of their body is troubling enough to them to undergo a significant surgical procedure. If at the end of their treatment they are no better, or they feel they are worse off, then they can feel even worse about themselves than before.

Refusing to treat patients

Some doctors reported that they refused to carry out cosmetic procedures on some patients, even though patients were keen to have the procedure done. Others noted that there were times when they should not have carried out procedures because they felt the patient's motivations were not sound. That is, they felt that patients were either not doing it for themselves alone or were doing it for the wrong reasons:

> This is the only branch of medicine that I've ever come across where the patient is absolutely upset, horrified that you don't operate.
>
> (Dr Green)

At the same time, doctors recognised that patients could be having cosmetic surgery for the wrong reason, or while they were in the wrong frame of mind. Some felt that it wasn't their role to refuse a procedure on these grounds alone or even ask why they were seeking the procedure in the first place:

> Well sometimes women come here and they're in a 'crisis', which is often when they seek cosmetic surgery. They've either had a divorce or a relationship breakdown or some other tragedy or grief or loss in their life – that sort of thing and ... they think if they have something done to themselves, it will make them better. Of course we know it is not true, but sometimes if you do bear in mind the cosmetic procedures ... it does improve their self-esteem and help

them in that way, so you can't very well say, 'Don't do anything'. You know? I think it is purely a personal choice what people want to do, so long as they're aware of the upside and the downside – it is their choice. Not mine.

(Dr Pears)

There were also varying approaches to the degree or amount of surgery that some doctors were prepared to carry out and it became clear through interviews, with both doctors and women, that doctors approach this issue in a number of ways:

I just say well I don't think the cost benefit or you know, the benefit versus the cost of it is going to be the right ratio for you. You're going to pay a thousand dollars to have ten cents worth of improvement and it isn't for me to do it for you. Then they go somewhere else.

(Dr Green)

There was general agreement among the doctors that most patients whom they refused to treat would go elsewhere and patients were often criticised for not accepting the doctor's judgement on whether a procedure was feasible or necessary:

When you go through the medical risks some say 'gee I didn't know you could die from an anaesthetic' or 'I didn't know that there would be scars' and they go away and never come back. A lot of them go to another surgeon who doesn't tell them those things and that's fine. That's between them and that surgeon.

(Dr Inch)

Doctors, like Dr Apple, referred patients to colleagues for a second opinion or if they felt that it was more appropriate that the colleague carried out the procedure:

if there is a medical reason the procedure is inappropriate. Or sometimes, when I think the person is … their body has a dysmorphic condition, which I think is not that uncommon and people are 'cosmetic junkies'. And they've had four or five procedures done elsewhere. And so actually, what I do with them actually, I know they're going to go elsewhere – so if not me, I try and get someone else to do the procedure.

(Dr Apple)

While these doctors talked of their discussions with women about the possible risks of surgery, it is clear that many of these risks, in particular those involving 'unrealistic expectations', are difficult to negotiate and not very tangible to women. Central to the argument about unrealistic expectations is the assumption that it is the doctor who is the arbiter of what is realistic and what is not. The woman is denied the opportunity to frame her own vision about what is realistic for her body.

Knowing when to stop

Criticism was levelled at some practitioners who carry out procedures with little attention to potential risks:

> I do the least possible amount of surgery to satisfy patients' needs. Younger surgeons do more and more procedures and increase the risks. But they don't tell patients and I have seen ghastly complications.
>
> (Dr Black)

Dr Cherry commented that the doctors' egos sometimes led them to carry out procedures that they should not attempt:

> But it is difficult to say no to a persistent patient – especially when you think you can help them. You know, I've had this discussion with a number of guys in this area and we do differ considerably, but I guess your ego gets in the way to a certain extent – in 'I can do this', but that doesn't necessarily mean that you should do it.

With the increase in demand for cosmetic surgery and the new technologies available, doctors who carry out cosmetic procedures will continue to work in an area that is in the twilight zone between what people regard as 'real surgery' and 'just cosmetic work'. While they may try to explain the very real medical risks that surgery brings with it, they are also dealing with the risks of aesthetic outcomes that do not meet the patient's expectations or wishes. At the same time, they are practising in an environment where medicine is meeting the market, where money is to be made through offering a service that changes the way people look rather than, necessarily, improving their health. Given these considerations, what are the similarities or differences in the way women understand risk in the cosmetic surgery and the way doctors understand it?

Conclusion

The data gathered from both doctors and patients interviewed for this study point to a difference in understanding between doctors and their patients. While women may not fully appreciate how the surgical process will affect them, doctors seem not to inform them about risk in a way that facilitates a thorough understanding of the issue. Doctors may outline major risks, but do so in such a way, and by using means, that dilutes the significance of the message. Thus, the use of computer programs and ancillary staff may not be as effective as would be the doctor going through the risks in detail and being able to answer questions posed by the patient that may arise during this process. At the same time, advertising that says that some procedures are 'safe, effective, same day' giving 'excellent results' as well as a lack of information about cosmetic surgery may mean that women do not foresee certain risk as being a part of the dynamic of cosmetic surgery. Visual messages, coupled with a lack of adequate knowledge, may give women an impression of something less clinical, more straightforward and less complicated than 'real surgery'. As Fraser (2003b: 86) notes:

> Central to the discourse of agency that saturates much cosmetic surgery discourse in women's magazines is the notion that the individual is able to evaluate the relative risks and rewards of undergoing cosmetic surgery, and, it is implied, that the individual has adequate access to information in order to make informed choices.

The post-operative recovery phase was another area where women felt they were kept in the dark. Again, perhaps the ethos surrounding cosmetic procedures mitigates against it being seen as something that can make you look and feel dreadful for some time. However, the effects of the surgery are no less real or severe, whatever the end result achieved. While women may again put up with unexpected post-operative pain and discomfort, they are less likely to do so if the other facets of their relationship with the doctor are unsatisfactory, and if they feel they have not been given adequate information.

There is significant evidence according to this study that doctors and patients involved in cosmetic surgery operate in different spheres of knowledge and understanding about each other, about the procedures they are involved in and the risks that these procedures may have. These spheres sometimes interlock but never seem to fit together exactly. As such, when each group seems to understand risk in very different ways, there is a greater likelihood that a lack of shared understanding will result in outcomes that are unsatisfactory to both parties, potentially physically and emotionally costly to women, and financially and professionally costly to doctors and their insurers.

Doctors fail to understand what risks women want to be informed of and they make decisions that affect women's bodies in ignorance of this. This means that once women choose to undergo cosmetic surgery, there is a steady erosion of agency over what happens to their bodies during the surgical process. The lack of information on risk is one significant element contributing to this erosion.

The most significant risk for women is that they do not get what they want and that they come out of surgery feeling worse about their body than they did before. There are outcomes that are clearly objectively poor and these are related to medical problems or mistakes that can be quantifiable and may be remedied. There are other outcomes that are measured subjectively and this is where the real problem lies. When women get an aesthetic outcome that is not of their choosing, it is very difficult, if not impossible, for them to have the surgery reversed and they are left with a body that they feel as bad about, if not worse, as they did before the surgery. It is a body imposed upon them rather than one of their own choosing.

9
Overview

So far this book has reported the experiences of women who have undergone cosmetic surgery and the interactions they had with doctors who carry out this surgery. It has also considered how doctors describe their role in the process of cosmetic surgery and their explanations of why they chose to engage in this work. This chapter will draw together the key themes discussed in previous chapters and highlight the often problematic and disparate discourses that abound in cosmetic surgery. The study on which this book is based focused on three key areas: what motivates women and doctors to take part in cosmetic surgery, how they communicate with each other and how they understand the risks of cosmetic surgery. It situates cosmetic surgery within the Australian context and the debates and deliberations that are ongoing in Australia about cosmetic surgery. However, although this study was conducted in Australia, these debates are common across Western countries and will become increasingly salient as demand for cosmetic surgery grows and cosmetic surgery techniques develop.

As we have seen, internationally there is a dearth of sociological research on cosmetic surgery, in particular, little empirical work has been undertaken on women and doctors and how they relate to each other in this setting. As discussed in Chapter 4 Davis's (1993; 1995; 1996; 1997a; 1998; 2002; 2003a; 2003b) work seeks to explain what motivates women to undergo cosmetic surgery and explains these motivations as being acts of agency and self-determination. Others (Bordo, 1993; Dull and West, 1991; Morgan, 1991) have challenged this, arguing that agency is compromised by a culture that has certain expectations of women and how they should look. The reasons women gave for wanting cosmetic surgery were detailed in Chapter 6 and were similar to those proffered by Davis. However, in this book, the debate moves beyond the motivations of women and includes the role and influence of the doctor in cosmetic surgical outcomes. In so doing, it develops the knowledge base about cosmetic surgery and its complexities. While it can be argued that women who undergo cosmetic surgery are initially independent agents and decision makers about their bodies, their abilities to maintain that agency and take their 'lives in hand' (Davis, 1995: 181) are heavily impacted, at all stages of the cosmetic surgery process, by the doctors who carry it out.

Women's reasons for wanting cosmetic surgery are situated in the self, but constructed through social and cultural influences (Bordo, 1993; Dull and West, 1991; Morgan, 1991). Women trust doctors with their bodies, doctors who are also exposed to social and cultural norms about women's bodies. During the deliberations about surgery, these influences come into play and are exposed. Thus, the way doctors interpret and reformulate women's reasons for wanting surgery, and the ways they communicate with them about that surgery, are delimited by their cultural experiences and understanding and through their position of power and authority as experts in that setting.

Again the understanding that women and doctors have about risk in cosmetic surgery differs. It is here that women and doctors often talk past each other and fail to be meaningfully engaged as cooperative partners who have a shared understanding of what women seek from cosmetic surgery, the risks they are prepared to take to achieve this and the risks they want to have outlined to them. This is an important element in cosmetic surgery, not least because there has been a significant rise in litigation in this area (Pirani, 2005), but also because full disclosure of risk is pivotal to giving fully informed consent to surgery.

Of course, not all doctors working in this field misunderstand women, their motivations and what they need to be reassured of. Nevertheless the issues of misunderstanding and miscommunication between doctor and patient are significant enough to be of concern and to adversely affect the outcomes women get from cosmetic surgery. Many women explained how communication breakdown and misunderstandings with their doctor had impacted them. While some reported having been treated very badly by their doctors, some of their experiences were simply extreme examples of what other women reported in this study. Other women were not totally happy with their outcomes or with their experience of cosmetic surgery, but had not complained to their doctors about that surgery. They reported that, as a whole, they felt they had been treated fairly and that the doctor had communicated adequately with them. In fact, women were prepared to put up with a great deal and take responsibility for the decisions they had made about their bodies as long as they felt they had been listened to and that the doctor had spent enough time with them.

In summary, any understanding of the phenomenon of cosmetic surgery needs to situate the practice within its cultural context, understand the complexities within the cosmetic surgery process, consider the nature of the interaction between women and their doctors and appreciate the risks involved in cosmetic surgery. This chapter will now provide an overview of the key findings of this study and situate them against the relevant literature reviewed in Chapters 2, 3 and 4. In order to understand why women choose cosmetic surgery in the first place, it is important to consider the influence of society and culture in addition to their individual wishes about their bodies.

Culture's influence on women's bodies

> Over the past hundred and fifty years, under the influence of a variety of cultural influences, the body has been forced to vacate its long-term residence on the nature side of the nature/culture duality and encouraged to take up residence, along with everything else that is human, within culture.
>
> (Bordo, 1993: 33)

The question of women, the body and body image has been discussed in Chapters 3 and 4. Women, over the centuries, have gone to great lengths to control their bodies for the sake of beauty. From foot binding to the wearing of a corset, historical accounts are replete with tales of women responding to social, cultural or religious pressures to adapt their bodies to meet current norms. From the beginning of the twentieth century, as women in the Western world became more visible in the public domain, the norms about beauty and acceptability became subtly and significantly integrated into women's consciousness across social and economic divisions.

The widespread availability of print and electronic media has enabled constant access to visual images. As part of this trend, versions of female beauty have been standardised, homogenised and normalised. Bordo (1993: 36) states that we are 'swaddled in culture'. Culture enfolds and disciplines the body and, particularly for women, makes insecurity and dissatisfaction the common parlance of any debate about how it feels to inhabit flesh that does not conform to that which is deemed to be appropriate.

A number of studies of the effects of media presentations of thin and beautiful women have been undertaken. A meta-analysis (Groesz et al., 2002) of 25 of these studies found that there was a significant negative effect on body image after study participants viewed thin models, compared to average-sized or large-sized models or inanimate objects. Gender differences in body image and body dissatisfaction have also been found to be increasing (Feingold and Mazzella, 1998).

Researchers on body image have recognised that while a significant number of studies have looked at adolescents and young adults and their body image, there has been little attention paid to middle- and older-age groups. A study by Paxton and Phythian (1999: 120) found that

> Females had significantly higher body disparagement, feeling fat, salience of weight/shape, appearance orientation than males across this age span ... These attitudes are in line with those encouraged within our society and reflected in the media.

Thus, women's negative feelings about their bodies are not only a function of adolescence or young adulthood, but are carried across the lifespan into their later years. Women's decisions to have cosmetic surgery, whatever their age, are socially situated and culturally influenced.

While satisfaction with one's body image has a psychosocial basis, and those who seek cosmetic surgery are thought to have poor body image and poor self-esteem, Sarwer et al. (1998) found that the cosmetic surgery patients interviewed were not dissatisfied with their overall appearance, compared with the norm. Their dissatisfaction lay with the body part that had received cosmetic surgery and this level of dissatisfaction was greater than a normative sample. Given the availability of cosmetic surgery and the prevalence of advertisements for cosmetic surgery in a variety of popular media, it is not surprising that women, at different times of their lives, will seek solutions to negative attitudes about parts of their bodies.

The narratives of women detailed in this book explain how cultural influences came to bear on them. Their need to change their bodies to fit into society suggests that there are effective social and cultural messages that articulate what is acceptable and what is not. Their active self-surveillance led them to want to minimise the surveillance of others. Or, to put it another way, they see that there is a visual homogenisation existing and want to be part of this rather than to resist it. Descriptions of hiding those parts of their body that did not fit in, even from those closest to them, demonstrate how intensely their feelings of 'otherness' affected them. Women who had lived with feelings of inadequacy for some time, and who had been publicly ridiculed about parts of their body in the past, sought either to distract attention by improving another part or by being less visible. They adopted culturally approved beauty practices through make up and dressed to distract attention, but felt they never quite succeeded and saw cosmetic surgery as the only real permanent alternative to a constant state of exposure.

Women who had inherited the feature they wanted to change felt that they were rejecting part of who they were by wanting that change. Disguising their ethnic heritage or their genetic inheritance meant that different meanings could be read into their decision to have cosmetic surgery. For them, the meaning was clear: they did not want to be noticed but felt that, for their family, this choice was not well understood. Those who were changing a feature that was an ethnic marker, such as their nose, were construed as exhibiting a cultural anxiety about themselves and their heritage by wanting to look ethnically neutral. Kaw (1998) explored the surgery women from Asian backgrounds underwent to look less Asian and more American. She saw these women as having internalised a gender ideology as well as a racial ideology about what they need to do to fit into contemporary American society. The message is consistent and is about blending in and feeling accepted by society rather than being different from it.

In this book, the stories of women whose dissatisfaction with their bodies had been brought about because they had become mothers or had lost weight, articulate a need to return to who they were. They felt that they no longer fitted in and that something had been taken from them even though they had taken on board all the messages about what they should rightly do as mothers and as healthy individuals. The women who had breastfed children were particularly adamant that doing the right thing for the health and well-being of their children should not rob them of their breasts and they wanted them back. Their dismay at having 'deflated balloons' or 'socks in the wind' was exacerbated because they had, on the one hand, been good mothers and conformed to social expectations of a what a 'good' mother should do but, on the other, they felt they had sacrificed their bodies.

Fitting in within acceptable boundaries was also a reality for women who felt that their ageing bodies disguised who they really were. For most of the women, it was not growing older that was the issue but their need not to look as if they were. It was clear that they did not want to look younger, but neither did they want to be defined by their age. Thus, needing to look better in their workplace was important so that they could continue to compete in that environment: an environment which clearly sees looking youthful as an important element if women are to succeed in particular workplaces. Some women felt they looked tired and angry and that this brought unfavourable reactions from others. Again, this did not reflect who they felt they were and brought unwanted attention from others. Here, and in a multiple ways, current cultural standards and expectations about women's bodies are subtly reinforced.

Cosmetic surgery inscribes culture on to women's bodies and 'transforms the material body into a sign of culture' (Balsamo, 1996: 58). It is through the surgeon and the surgical process that an idealised version of female beauty is achieved by inscribing on women's bodies a certain type of physical appearance. Thus:

> Cosmetic surgery is not simply a discursive site for the 'construction of images of women', but a material site at which the physical female body is surgically dissected, stretched, carved, and reconstructed according to cultural and eminently ideological standards of physical appearance.
>
> (Ibid.)

Spitzack (1987; 1988a) also sees cosmetic surgery as a means of cultural control. Spitzack proposes that there are three interconnected mechanisms that enable this. These are surveillance, confession and inscription. As already discussed, women in this study often felt as if they were under the surveillance of others. Their confession to the cosmetic surgeon led to the inscription of a more culturally acceptable form on to their bodies. However, the cosmetic surgeon's gaze is a disciplinary gaze, supported by medical power and knowledge that identifies the female body as flawed and in need of correction. The woman's acceptance of this analysis of her body is a confession, an admission that her body is 'diseased' and that the surgeon can inscribe beauty on to her body through his scalpel.

Gilman (1999) discusses the historical development of aesthetic surgery as a means through which people could 'pass' in society and fit in with socially and culturally accepted standards. Plastic and aesthetic surgeons became the means through which people could obscure signs that made them appear different while, at the same time, enabling them to belong in the world they inhabited. For women, cosmetic surgery and its practitioners have become tools that can help them pass more comfortably into a culturally acceptable femininity. It also fulfils the need not to be observed, not to be looked at, not to be surveilled. There is a significant irony in this because women have to put themselves forward for intense and close surveillance by the doctor if they want cosmetic surgery. This surveillance is not only by the doctor's gaze but also by the technological gaze of the tools of cosmetic surgery that visualise the body, reform it and reconstruct it.

Fitting in

In Chapter 3, the increasingly public exposure of the female body over the past century was discussed. Surveillance of the female body and its fetishization began during the mid-nineteenth century (Stratton, 1996). This led to the female body being 'spectacularly surveilled' (ibid.: 87). Aiding this surveillance was the development of film and photography which meant that female bodies could be 'captured' and circulated. At this time, women were increasingly present on the street without being accompanied by a man but their presence there was still on male terms, that is, their behaviour and demeanour needed to fit in with male expectations. With standardisation of look and dress, the normalising and disciplinary power of the male gaze reinforced the authority of men, both within and outside, the home (Foucault, 1979; Hillier, 1997).

As women's bodies have become more public, the homogeneity of look and fashion in society has become more pronounced. Brush (1998: 35) claims that

> The power of the normalizing judgement allows the registration of difference only as degrees of difference from the norm which is the ideal. The ideal of cosmetic surgery sets up an ideal face, a series of mathematical measurements and aesthetic ideals which construct the normal/ideal and influence the work of the surgeon.

Many women in this study reported that the outcome of their surgery was, indeed, more influenced by their surgeon than their own desires for their body.

Of course, to say that women are subject only to external surveillance ignores the very real self-surveillance that women practise every day on themselves. Foucault (1979) argued that those who are subjected to the continuous surveillance of others (in that case the prison guard) become so used to it that they internalise the gaze to such an extent that they constantly scrutinise themselves. Grosz (1994) and Bartky (1998) reflect on Foucault's contentions. However, they argue that he:

> treats the body throughout as if it were one, as if the bodily experience of men and women did not differ and as if men and women bore the same relationship to the characteristic institutions of modern life.
>
> (Bartky, 1998: 27)

Added to their self-surveillance is the medical surveillance that women are subject to throughout their lives (Oakley, 1998). Thus, women are used to internal and external surveillance that is entangled in cultural expectations of women's bodies functioning both in a reproductive capacity and as objectified entities. Women are accustomed to 'having things done' to their bodies without overt coercion but for, what is claimed by medical experts to be, therapeutic reasons. This 'colonization' (Morgan, 1991) of the body by experts is an ever-present part of women's lives. Thus, to believe that the act of cosmetic surgery is one of self-determination ignores the fundamental effects of culturally bound normalisation pressures and medically based surveillance techniques that are constantly present for women. The need to adhere to the 'normal' is, therefore, a strong motivation for women who feel that their bodies fall outside what is projected to them as normal through social and cultural structures and tools.

While all of the women's narratives in this study are concerned with the wish to be normal, what they do not elucidate is what 'normal' is. In wanting to be normal, women also emphasised that their desire was for a natural look. Achieving a natural look was seen as important so that they did stand out. But again, 'natural' was not clearly defined. The repertoires of 'natural' and 'normal' were also part of the doctor's narratives about cosmetic surgery.

All the women clearly explained that they wanted cosmetic surgery for themselves. However, they were also able to reflect on the cultural pressures that influence the way women's bodies are represented in society in general and particularly through the media. They understood that in the media, women are portrayed in a fairly homogenised form, but were still clear that it was for themselves that they wanted surgery and not because they wanted to conform to some social stereotype of a beautiful woman. Self-esteem and 'fitting in' does not occur in a social vacuum but is reflected through the ways others view us. The women in this study were doing it for the purposes of their own self-esteem but also as part of negotiating their place in a culture that focuses on the viewed body as a signal of compliance to certain cultural norms and expectations. In particular those women who wanted cosmetic surgery so that they could compete more effectively in their workplace were clear examples of the negative attitudes about ageing in society. To be at ease with themselves, and in response to such attitudes, they needed to address a part of their body that trapped them in a cycle of self-criticism, self-consciousness and the perceived surveillance of others. If their motivation was to be less obvious and to blend more with what they felt were more acceptable standards of aesthetic acceptability, to be 'normal', then this had to be communicated to and negotiated with their doctor.

Fraser (2001, 2003b) discusses the way in which nature or the natural is employed in the popular press, medical literature and within feminist critiques of cosmetic surgery. In the popular press and medical literature, natural and normal are often used interchangeably to make cosmetic surgery seem acceptable and usual. They are also used to delineate the preferred outcome of cosmetic surgery and to depict it as a means of achieving what nature has not provided. In women's descriptions of what they wanted from cosmetic surgery, being normal by achieving a natural look through cosmetic surgery was a consistent theme. These women also saw nature as having taken something from them, either through genetics, childbearing or ageing. Thus women saw that cosmetic surgery was a way of reclaiming what nature deprived them of. For instance, breast implants were not seen as an unnatural presence in their body but as a sign of the naturalness of full breasts for women. But implants that were too large were not natural, felt heavy and drew unwanted attention, thus placing these women outside the boundaries of what they felt was normal.

Fraser (2003b: 65) proposes that

> it is possible to argue that traditional constructions of nature help make possible the scientific/medical field of cosmetic surgery

Nature has always been seen as gendered and related to the female and the feminine which in turn make women the obvious candidates for cosmetic surgery. Fraser (ibid.) goes on to point out that

> one understanding of nature has been that it constitutes the inert raw materials on which culture sets to work. This view of nature sees it as something to be overcome, transcended and, in the process, moulded, adapted and controlled. It is a view that privileges culture over nature, and sees human existence as necessarily a struggle against the primitive impoverished forces of nature.

If this is the case, women in this study on the one hand struggle against nature and its consequences on their bodies, while on the other they feel the need to conform to the culturally defined ideal of the natural.

Nature and the natural are also terms that pervade medical texts on cosmetic surgery (Fraser, 2003b). In some of these texts nature is used as a reassuring presence to normalise cosmetic surgery. In conceptualising it in this way cosmetic surgery is seen as both more acceptable and accessible. This was also a rationalisation echoed in a number of ways in the doctors' discussion of cosmetic surgery. They agreed that women had cosmetic surgery because they wanted to look normal. However, these doctors were clear that they wanted to operate only on normal women, not on women whom they defined as having expectations that were outside what *they* thought was normal, or who *they* thought were having cosmetic surgery for the wrong reasons. There is an inconsistency in these accounts: on the one hand doctors want to operate on 'normal' 'healthy' and 'stable' women; on the other, they diagnose women as being psychologically affected by a 'flaw' in their body. Doctors saw women as psychologically burdened by the abnormalities in their bodies and regarded surgery as a curative option for this psychological distress. It could make the body and mind normal. However, they were very clear that they did not want to operate on women whom they regarded as 'dysmorphic', that is, as having a pathologically unrealistic view of their body. Miller et al. (2000: 356) argue that if cosmetic surgery belongs 'within the proper domain of medicine' then to serve the goals of medicine it should be treating those that are most affected by a condition rather than the least complicated or problematic cases. The narratives of doctors in this study suggest that this link between cosmetic surgery and psychological health does not extend to everyone who wishes to have cosmetic surgery. It only applies to those that they see as deserving and psychologically stable.

Psychology and cosmetic surgery

Haiken (1997) traces the development of the link between cosmetic surgery and mental well-being from the 1920s onwards. She identifies the creation of 'inferiority complex' and 'self-esteem' as psychological terms that enabled cosmetic surgery to justify itself as a tool for improving mental health. She states that

> By the late 1930s ... surgeons were in almost universal agreement about psychology's relevance to their work and proud of themselves for the progress they had made in incorporating these new ideas into their practice.
>
> (Haiken, 1997: 123)

As cosmetic work became more lucrative and in demand, surgeons reclassified certain physical conditions as deformities or disfigurements because this:

made it easier for surgeons who saw their function as dealing with serious problems to justify their work, to themselves, to others, and to official bodies.

(ibid.: 122)

Through linking the cultural weight given to physical appearance with psychological well-being, cosmetic surgery was now able to legitimise itself both in the eyes of the public and in its own eyes. Little has changed in this respect over the last eighty or more years, but the cultural grasp over the norms of bodily style and significance has intensified.

As discussed earlier, women's bodies have historically been seen as problematic and as an appropriate site for medical intervention. Affluent women were frail while poorer women, like poor men, were often characterised as base and not quite human (Weitz, 2003). Gynaecological surgery on middle-class women became an epidemic in the more advanced Western societies during the late nineteenth century, with doctors removing many healthy reproductive organs to treat a variety of physical and psychological symptoms. Such surgery was used to manage women who were disaffected or unruly. Social control through the scalpel evolved as doctors took over the management of other reproductive functions, such as childbirth, and further defined women's bodies as a site for regular medical surveillance (Cahill, 2000). Given this historical precedent, cosmetic surgery can be seen as yet another way of controlling women's bodies through medical intervention, while at the same time explaining it as an act of psychological necessity.

As we have seen, some doctors in this study explained their interventions as being psychologically beneficial for women while, at the same time, being careful to avoid operating on women whom they see as psychologically damaged in some way. Of course their views on women's psychological health are purely subjective given that they do not refer women for psychological assessment. The narratives of women, while attesting to the distress they had felt about parts of their bodies, did not suggest that they were, or that they felt, particularly psychologically unstable.

Doctor as artist and sculptor

Doctors in this study discussed why they were attracted to the discipline of plastic surgery. Some explained that, for those who had an artistic temperament, the discipline offered them the opportunity to satisfy their creativity. Haiken (1997: 221) in her history of cosmetic surgery states that cosmetic surgeons view themselves as 'artists, veritable sculptors in human flesh'. However, she goes on to state that these same surgeons do not see their artistic inspiration in any way influenced by culture or bound by it. Haiken disputes this saying:

> The mantle of artistry has allowed surgeons to claim a disinterested position – a position outside culture. Their standards, their values, and their artistic sensibility, they say, derive from the same timeless canon of craft, skill, and beauty that produced David and the Mona Lisa. Always suspect, this claim is simply no longer supportable: their own history demonstrates that American plastic surgeons are both products and producers not only of a culture of medicine but of a culture that is unique to modern America.
>
> (ibid.)

Art over the centuries has displayed ideals of beauty that were prevalent at a particular time and in a particular culture. Mona Lisa was a beauty of her time in the same way as a contemporary Miss Universe is considered a beauty of our time. Neither art nor beauty can be separated from cultural and social influences. If surgeons are artists, they cannot be immune from the same influences that pertain to physical beauty that affect us all. For women, the consequence of these influences is that they feel the need to change some part of their body so as to conform to a culturally acceptable standard. Doctors, in turn, interpret women's wishes through the lens of their particular, culturally defined version of that standard.

Brush (1998: 30) argues that cosmetic surgery is also about characterising the 'natural' bodies of women as fundamentally defective. She says:

> Cosmetic surgery involves an aesthetic judgement, a normalising gaze which divides the body into component parts and transforms difference into a closer approximation of the norm. The aesthetic practice of cosmetic surgery does not simply rely on refining the 'natural' body ... but works by re-defining standards of beauty and inscribing those standards on to the defective 'natural' bodies of women who fail to resemble, closely enough, the norm.

Doctors do not merely replicate and reflect cultural norms but redefine and reformulate them through their own views of what is culturally appropriate for women's bodies. At the same time they adopt medical language and terminology to translate bodily characteristics into symptoms and abnormalities. Thus, 'deformities' are identified and:

> the rules of symmetry and 'classical art theory' can be inscribed on to the unsatisfactory female body.
>
> (Brush, 1998: 30)

Through making cosmetic surgery a medical intervention rather than a personal beauty choice, it is redefined as a need rather than an option for women.

Medical and popular publications in cosmetic surgery identify the artistry of cosmetic surgery (Fraser, 2003b). In the medical literature the surgeon is depicted as both artist and expert technician, with the successful surgeon being accomplished in both domains. Thus 'successful cosmetic surgery becomes a matter of talent, requiring an indefinable "touch" like that of the master artist' (ibid.: 135). The creativity and sensitivity of artist is linked to the technical and scientific skill of the expert which, in turn:

> works to replace images of operating theatres and surgical instruments with rather more romantic images of art, in which the female body is the artwork in a very traditional sense.
>
> (Ibid.)

In this way, cosmetic surgery is not only elevated to an artistic endeavour but is also separated from any visual identification with cutting and bleeding that is associated with operating theatres and surgical tools. Such images are soothing and reassuring for women who decide to seek cosmetic surgery, but they camouflage the reality of that surgery and they fail to fully account for the creative role the doctor plays in cosmetic surgery.

The outcomes of cosmetic surgery are mediated by a range of complex yet interconnected factors which include the doctor's artistic input, cultural influences on doctors and their female patients, and the medical authority and power that is vested in the doctor. Fraser (2003b: 135) equates the female body to 'raw material' that the doctor/sculptor can work on and:

> The metaphor of the artist shaping the unrefined clay of the surgical body not only convincingly links cosmetic surgery with femininity on a profound level through the relationship between art object and

femininity within the sexual imagery, but it also reproduces traditional elements of passivity, and acceptance of the mastering gaze
and touch.

In this research, the most striking example of this is women's experiences of breast implants. There was only one woman who had implants
that were smaller than she wanted. Lorna wanted 'big and obscene'
but had settled for a 'D' cup. Every other woman felt that the implants
were larger than they had wanted or expected. Given that research
indicates that men's preferred breast size is larger than women's (Huon
et al., 1990), this suggests that doctors give women breasts that please
their own notions of attractive breast size and proportion rather than
what the women want. Women commented that their new breasts now
attracted the attention of others because they were too prominent and,
while one woman was so unhappy with the outcome that she complained about her treatment, others just accepted the result they had
been given.

The case of rhinoplasty was another example where women described
the outcome of surgery as better than what they had before, but in reality the aesthetic outcome was not decided by them. The final form of
the nose was very much left to the doctor to create and, as one doctor
put it:

> Nefertiti is a stylised version of female beauty, long jaw line, elegant
> neck, and the nose I've been creating for years is her nose.
>
> (Dr Black)

Women also discussed the way doctors suggested other procedures that
may complement and enhance the results of the procedure they were
seeking. This was particularly true for facial surgery. Here, doctors diagnose women as having a different, or additional, aesthetic problem to
the one women thought they had. This can make women feel that they
are even more flawed than they thought they were and can make them
relinquish the decision about the nature and extent of the changes
made to their body to the doctor. The doctor's eye and expertise are a
persuasive force for women who already feel they are flawed and, their
influence is such, that women are likely to accept their judgement.

Spitack (1988a) describes her own experience in a cosmetic surgery
clinic where she was convinced by the surgeon that she needed to have
her skin resurfaced as well as a rhinoplasty. She describes her feelings
of being externally scrutinised and exploited by her surgeon and of her

self-scrutiny and vulnerability to the surgeon's reassurances that he could solve her 'problems'. Spitzack's description of internalising the doctors aesthetic judgement of her mirrors the experiences detailed by women in this book. These are judgements that are heavily laced with the surgeon's power over the women and his influence on how their bodies should look.

The lack of shared understanding between doctors and women

Sharing an understanding of why women wanted cosmetic surgery and what they wanted from it is at the heart of the cosmetic surgical exchange. It was significant in this study that doctors and women used different language to describe the same features. Thus, women described what bothered them about their bodies as flaws and imperfections while doctors were more likely to describe the same feature as a deformity or abnormality. In so doing doctors offer a medicalised diagnosis and, as Balsamo (1996: 63) states when discussing ageing and cosmetic surgery:

> This is a simple example of the way in which 'natural' charac-
> teristics of the aging body are redefined as 'symptoms', with the
> consequences that cosmetic surgery is rhetorically constructed as a
> medical procedure with the power to 'cure' or 'correct' such physical
> deformities.

This redefinition and reconceptualisation is not confined to the ageing body, it is also used to describe features such as breasts and noses. Thus, women with small or sagging breasts were diagnosed as having an abnormality that should be medically treated. The use of such language enables cosmetic surgery to sit more comfortably as a medical problem and, in this way, can be more closely aligned to plastic and reconstructive surgery and identified less as a consumer commodity. This, of course, is comforting to those who carry out the procedures. As we saw in Chapter 5, there is a real and unresolved tension within cosmetic surgery as to who should rightfully practise it. There is also an acknowledgement from those who do practise cosmetic surgery that the medical profession, in general, sees cosmetic surgery as a 'borderline' medical practice (Miller et al., 2000) which intervenes in the operation of the body where there is no real medical necessity to do so. But again, and as mentioned earlier, doctors who practise cosmetic surgery identify the need for these interventions as psychological, rather than physical, and, as such, reframe what they do as necessary medical intervention.

The 'psychology with a scalpel' argument is a double-edged sword for these practitioners. On the one hand they argue that they are 'changing the body to cure the soul' (Gilman, 1998) but on the other there is silence about the psychological consequences to patients when things go wrong. In cosmetic surgery, and as reported in the narratives of women in this study, getting an unsatisfactory outcome can be psychologically and physically devastating for patients. This issue will be discussed more fully later in this chapter. The way doctors try to avoid this is partly through patient selection and this is often where they are at odds with some of the patients they see.

Women reported relatively straightforward reasons for having cosmetic surgery: that is, they wanted to look and feel 'normal' and wanted to 'fit in'. While it could be argued that they felt emotionally at odds with their body, or part of it, they did not report any obvious significant psychological condition or illness. They wanted cosmetic surgery for themselves and not to please a partner. Doctors' explanations of why women wanted cosmetic surgery were different to those proffered by women. For instance, doctors were at pains to explain that there were good and not-so-good candidates for cosmetic surgery. The crux of this was that a good candidate did not have 'unrealistic expectations' of what could be achieved through surgery. Also, while some doctors felt that some women were psychologically unsuitable for surgery, they did not refer them for psychological assessments. To reiterate, these two issues are intertwined, yet contradictory. On the one hand, it is only psychologically 'stable' women who are deemed suitable for surgery, but on the other they are seen to be psychologically affected by their flawed bodies to such an extent that only cosmetic surgery can 'cure' them. Those who did not agree with their doctor's view of the outcome of surgery, and were not 'docile' recipients of his evaluation of their body and how it should look, were judged to be unsuitable for surgery on account of their 'unrealistic expectations'.

If we think of a 'realistic' expectation as achievable, safe and conservative, then women's narratives attest to the fact that this is what they wanted. To return again to their need to fit in, the women felt that this could be achieved by doing as little as possible, that is, by having 'C' cup breasts or a facelift that made them look 'tidier', not younger. What they often got, however, were breasts that were larger than they wanted, a nose that they had not actually chosen and a facelift that was too severe. What this suggests is that the discourse surrounding and constituting cosmetic surgery is not a unitary concept and is not something shared, or understood in the same way, by doctors and the women they operate on. It also suggests that the arguments put forward by Davis (Davis, 1993, 1995), that women are exercising agency in choosing cosmetic surgery, are flawed. Fraser (2003a, 2003b) argues that

> The repertoire of agency presents women with a dilemma in terms of self-determination and reliance on medical expertise. In some ways this is a reflection of traditional dilemmas women face about 'being oneself' as the best means of gaining masculine approval. Here, in a thoroughly paternal medical context, agency is encouraged only so far. Desirable when it disregards the opinions of friends, family or lovers and when it funnels ambition and professional commitment into appearance, it must never be allowed to hamper the influence of surgeons over their patients.
>
> (Fraser, 2003a: 40)

There is little doubt about the influence doctors had over the outcome of surgery amongst women interviewed for this book. Many women accepted this input while others saw it as having taken away their autonomy over their bodies. In retrospect, many felt that, either they had not adequately communicated what they wanted, or their doctors just hadn't listened to them.

Explanations by women of why they wanted cosmetic surgery were often inconsistent with the explanations proffered by doctors. Women's rationale centred on wanting to fit in with what they saw as normal in their environment. Doctors, on the other hand, described women's need for cosmetic surgery as a psychological intervention achieved through physical change. While there was an acceptance on the part of women that cultural influences do have a part to play in the popularity of cosmetic surgery, they saw themselves as wanting limited intervention to enable them to operate in *their* world, rather than in the world depicted in the popular media. Doctors saw themselves, not as technicians but as artists whose role is to create beauty or normality in women. This suggests that the surgeon's creativity can override the patient's aesthetic wishes. It also suggests the 'mastering gaze and touch' (Fraser, 2003b: 135) of the surgeon and the passivity of the woman as the artist's canvas. Here, women's agency over their bodies is subservient to the surgeon's creativity. Thus, rather than achieving what *they* wanted from cosmetic surgery, women too often felt that their motivations for surgery were misunderstood and that the outcomes of surgery reflected this.

Communication between doctors and women

If women are to achieve satisfactory outcomes from cosmetic surgery, they need to be able to communicate to their doctors what they want and why they want it. Conversely, the doctors must be able to appreciate women's reasons for cosmetic surgery and understand what aesthetic results they are seeking. As we have already seen, there were significant disparities between what women reported as their reasons for seeking surgery and what doctors understood these reasons to be. In this study, women spoke about doctors who did not understand why they wanted surgery and what they wanted from it. The doctor's narratives also suggested that they viewed women's motivations for surgery from a particular perspective that could be said to reflect their own objectification of women's bodies. As we have seen, doctors were keen to ensure that women were 'doing it for themselves' and did not have 'unrealistic expectations'. Doctors' accounts also describe how they saw some women as more deserving of cosmetic intervention than others. Such repertoires of self-determination, independence and being deserving were also present in women's narratives. Davis (1995) and Gimlin (2000) both explain women's decisions to have cosmetic surgery as being internally driven. Even if there is a shared understanding of *why* women want cosmetic surgery, this does not automatically translate into an agreed outcome of that surgery. We have seen that the process of negotiating the outcome of cosmetic surgery is complex and dynamic. Fundamentally, it is a process where communication between doctor and patient is pivotal to the end result. The less tangible visions that doctors have about the female body and that women have about their own bodies also hold significant sway. In this study, the schisms and contradictions between the women's and the doctors' accounts of this communication indicate that there is often a lack of shared understanding about various aspects of the process of cosmetic surgery.

The question of successful doctor-patient communication has, not surprisingly, produced a significant body of literature. Some of this literature will be reviewed later in this chapter. First of all, however, the specific case of communication in the cosmetic surgery setting needs to be explored, as well as how the particular forms of that communication impact on, and influence, the outcome of cosmetic surgery for women.

Koch et al. (1998: 197) suggest that 'the majority of lawsuits arising from cosmetic surgery cases are based on mis-communication between the physician and the patient'. This is not surprising given the complex and diverse communication methods utilised in cosmetic surgery. Usual doctor-patient interaction is focused on the consultation. In cosmetic surgery there are a number of other players involved, as well as a range of other techniques. Even before the patient enters the clinic, significant communication concerning cosmetic procedures has been entered into and it is this aspect of communicating with patients that will be discussed next.

Attracting potential patients

Advertisements and publicity surrounding cosmetic surgery are important for drawing women's attention to procedures, and to the doctors that can carry them out. Even if doctors themselves do not advertise directly, the publicity surrounding cosmetic surgery raises awareness of what can be done. Fraser (2003b) contrasts the way in which advertising is either seen as making women susceptible to the trappings of cosmetic surgery, or as providing information for women that can help them make choices about cosmetic surgery. This is a reflection of the 'dupe' versus 'agent' debate among many feminist thinkers. Many women in this study chose their doctor because of publicity, either through direct advertising, magazine articles or advertorials about cosmetic surgery and/or the cosmetic surgeon. There is no doubt that advertising does have the effect of making cosmetic surgery more visible, promoting it as accessible and introducing practitioners to clients. However, as Fraser (2003b: 123) also points out, the boundaries of advertising and medical material are 'broad and fluid' so that

> For the lay reader, distinguishing between sound medical advice and sales material can prove difficult.

It is here that the nub of the problem really lies. Consumers are lured into trusting that their bodies can be moulded scientifically to be what they want it to be. The actuality of scalpels and the complexity of changing the body to fit what *they* want is never addressed. It should be no surprise then that women feel that they can achieve what they want from cosmetic surgery, rather than having to settle for what the their doctor can give them.

Since the end of the 1990s in Australia there has been sporadic interest and debate about the ethical, moral and legal position of advertising cosmetic surgery. In particular, the use of photographs to show bodies before and after surgery has been criticised. The Cosmetic Surgery Inquiry undertaken in New South Wales (New South Wales. Health Care Complaints Commission and Cornwall, 1999) reported that claims made about cosmetic surgery through 'before and after' photos 'may be misleading and 'deceptive' (ibid.: vii). In their deliberations on advertising during the HCCC Inquiry, those representing the consumers categorised women as susceptible to the promises implied through advertising. Doctors, on the other hand, were more likely to see advertising as providing initial information to women about cosmetic surgery, with any misconceptions about what cosmetic surgery can do being corrected in consultation with the doctor.

Describing women as open to exploitation through advertising, does categorise them as victims and ignores the trust we all put into medicine and the medical profession. While we may be suspicious of much advertising, claims made about doctors, their skills and how they can use these skills to change our bodies are less likely to make people suspicious. Women have no reason to believe that the before and after photographs shown in these advertisements, or any other visual material, are anything but accurate and honest. They put the same trust into technologies used in the clinic by the doctor and his staff and, while the use of before and after photographs may be regulated in advertising materials, they are still used within the clinic. Such materials, together with computer-based imaging technology, are used regularly to describe the types of outcomes women can achieve in cosmetic surgery.

Visual technologies

Computer imaging is promoted as a tool whereby doctors and their patients can jointly visualise an outcome of surgery. This situates such imaging as value free and independent of human input. The use of computers removes 'the final judgment' from the individual and gives it instead to the computer, as an independent arbiter of beauty (Spitzack, 1988b: 11). This enables women to believe that what the computer shows is how they should, and could, really look. However, Balsamo (1996: 58) maintains that

> Cosmetic surgeons use technological imaging devices to reconstruct the female body as a signifier of ideal feminine beauty ... The technological gaze refashions the material body to reconstruct it in keeping with culturally determined ideals of Western feminine beauty.

Through technological, computer-assisted communication, the 'ideal' female form is negotiated. The power of computer imaging to convince women to believe that they could look a certain way was confirmed by the reports of some women in this study. But there were mixed reactions as to the efficacy of the tool from both women and doctors. Some doctors felt that it offered the opportunity to achieve an agreed understanding of what outcome could be reached, others felt it was not any more useful that pencil and paper. In many ways, computer imaging was used to compare women's current (flawed) feature to a new (improved) feature. While this gave women some idea of what they could look like after surgery, it also cemented the notion that they were defective and needed, through surgical intervention, to be made whole again.

The consultation and cosmetic surgery

Most of the research and education in doctor–patient communication and in the medical consultation has focused on primary care. Although little work has been done with specialist practitioners, the models of communication are appropriate across areas of medical practice and it is useful to outline them here.

The literature on doctor–patient communication underlines four models (Emanuel and Emanuel, 1999):

- Paternalistic: In this model the doctor acts as the patient's guardian, implementing what he/she feels is in the patient's best interests.
- Informative: The doctor provides the patient with all the relevant facts and the patient determines the outcome.
- Interpretive: The doctor elucidates the patient's value system and assists in selecting the best intervention.
- Deliberative: Here the doctor acts as a friend or teacher to help the patient decide on the best intervention.

The doctor-centred paternalistic model vests the power in the consultation in the hands of the doctor while, at the other end of the spectrum, the deliberative model places the power and decision making in the hands of the patient. The way doctors communicate with their patients has a direct impact on patients' health and well-being. In particular a greater participation by patients in the consultation improves patient satisfaction and compliance with treatment (Emanuel and Emanuel, 1999).

Kirkland and Tong (1996: 157) argue that the deliberative model should be the only model operating in the cosmetic surgery clinic. However, they acknowledge that

[d]eliberative cosmetic surgeons are relatively rare.

This is not surprising given the power and authority assumed by, and vested in, specialist medical practitioners. Rose (1996) sees the 'authoritative' discourse of professions like medicine and law as being central to, and convincing in, the regulation of much human action. Although cosmetic surgeons, as specialists, occupy the more powerful position in the doctor–patient relationship, this is really not unusual in the medical context. While the literature clearly indicates that successful communication is imperative to the health and well-being of those who seek medical help, it also shows that doctors often have significant problems in communicating effectively with patients. Indeed, communication, or lack of it, is at the root of most malpractice claims (McClean, 1989). In assessing what we know about doctor–patient communication, Simpson et al. (1991: 1385) state that

> [m]ost complaints by the public about physicians deal not with clinical competency problems, but with communication problems, and the majority of malpractice allegations arise from communication errors ... a high proportion of patients do not understand or remember what their physicians tell them about diagnosis and treatment.

Ong et al. (1995: 903) note that 'inter-personal communication is the primary tool by which the physician and patient exchange information'. Unfortunately it is here that the relationship often fails, and patient outcomes and satisfaction suffer. They outline the purposes of medical communication as:

- Creating a good interpersonal relationship;
- Exchanging information; and
- Medical decision making.

One indicator of dissatisfaction with their doctors, reported by a number of women in this study, was that they felt that they had not established rapport with them. Some women talked about having a 'gut feeling' about their doctor and his communication style, but had explained this to themselves as being 'just how doctors are'. Women who had visited more than one doctor compared their styles and talked of not being comfortable with some but feeling at ease with others. In order to establish a relationship with their doctor, patients must be able to spend time with them. Many women had only one consultation and some of these were as short as seven minutes. Others spent some time with the doctor, but also with ancillary staff such as nurses and practice administrators. For women, however, it was the time spent with the doctor that was most important. Women who were the most dissatisfied with their treatment reported that communication between them and their doctors had been poor and that they had not been able to build a trusting relationship with them. Other women had not been particularly happy with the outcomes of their surgery, but were happy with the communication they had with their doctor. They felt that the doctors had spent time with them, listened to them and that they had established a rapport with them. This meant that they accepted their outcomes and did not feel the need to complain about them.

Being able to exchange information depends on the time spent with the doctor as well as the ability of patient and doctor to understand each other. Studies have shown (Ong et al., 1995) that there can be a lack of concordance between the perceptions of the doctor and that of the patient. Thus, doctors' ways of communicating information can often ignore the personal relevance and weight of certain information to patients. In cosmetic surgery, there can often be fundamental misunderstandings between doctor and patients. For example, Dora realised, in retrospect, that she and her doctor were at 'cross purposes' and her doctor had not understood what she wanted from her surgery, nor why she wanted it. She felt that this had resulted in an outcome that was not at all what she wanted or expected.

The flow of information between doctor and patient is rarely value free and is usually controlled by the doctor. In particular, there is well-documented evidence that there are significant differences in the ways women and men communicate. Where cosmetic surgery is concerned, the communication process also needs to elicit from the patient what her expectation is regarding the outcome of the procedure and if the practitioner is able to meet it.

Much of the early work in this area looked at medical consultation as an interaction managed by the doctor around a structured set of tasks, including the taking of a medical history, examination, diagnosis and treatment. Indeed, the consultation is 'task oriented', in that it needs to move through distinct stages. Davis (1988: 52), in an overview of studies in the 1980s, outlines these issues but also discusses how these studies describe 'management problems' in the medical interview. The problems that arose were analysed from the doctor's perspective and the doctor and patient were seen as 'peers in the interaction game, engaged in the mutual and cooperative production of an "orderly" and "accountable" social world'. Such an analysis ignores the fact that patients and doctors are not equal players in the interaction.

The extent and power of medical knowledge and authority was discussed in Chapter 4 and this is particularly significant for women whose bodies and bodily functions have been put under increasing surveillance and analysed systematically, and regularly, by the medical profession since medicine established itself as a distinct discipline. In the cosmetic surgery clinic, the superior knowledge of the doctor will have the most powerful influence on women seeking cosmetic surgery. If we remember that many women access cosmetic surgery in secret, without telling any friends or family members, their dependence on the doctor as friend, confidante and ally is more marked. Women in this study valued the time doctors spent with them and were more at ease with doctors who were prepared to spend some time with them. They were forgiving of doctors who may not have given them exactly what they wanted, if those doctors had established a rapport with them. On the other hand, doctors who had not handled the consultation well, or who had been arrogant or dismissive were more likely to be the subject of complaints by women who felt betrayed on a number of fronts in the cosmetic surgery process.

Women also spoke of the ancillary staff in the clinic who dealt with various aspects of cosmetic surgery consultation. This is, again, where cosmetic surgery as a medical encounter meets cosmetic surgery as a commercial exchange. There is no doubt that it is more profitable for doctors to employ these staff so as to free themselves up to see more patients, otherwise their employment would make no business sense. However, they remove the patient from direct contact with the doctor. Women talked of the frustration of not being able to ask questions of the doctors and of their being answered by the practice staff instead. While some of these staff would be nurses and have some medical knowledge, others were receptionists and administrative staff. The Cosmetic Surgery Report (Health Complaints Commission, 1999: 4) notes that

> Nurses also play a significant role in the cosmetic surgery industry. In addition to their traditional roles, nurses also perform some cosmetic procedures and provide patient counselling in some plastic and cosmetic surgery clinics.

The women in this study were very clear that, for any medical information or intervention, it was the doctor they wanted to see and speak to and not practice staff. Both women and doctors mentioned that the explanation of the potential risks of procedures and the signing of consent forms was frequently undertaken by practice staff. For women, the implication of what they were consenting to was sometimes not clear, and the choices they were making about their bodies were often not choices informed by good, or adequate, information. Davis (1995: 117) says:

> Choice presupposes that the individual has a viable option to choose from. Informed consent assumes that she has sufficient information to understand and evaluate the intervention and that her consent has been freely given: i.e. not coerced.

If women do not fully appreciate the consequences of their choices then their agency over their bodies is significantly constrained. If the choice is between continuing to feel the way they do about their body or, potentially, feeling worse because the outcome of their surgery was not what they wanted, would women make the choice to have surgery?

Davis (1995) argues that to see women as incapable of making their own choices and providing informed consent for the cosmetic procedures they undergo is to deny their competency and their ability to make decisions about their bodies. However, she concedes that choices about cosmetic surgery are not made 'under conditions of perfect knowledge' (ibid.: 119). We have seen in this study that women often have insufficient information, are not able to spend adequate time with their doctor and rely on practice staff to provide information to them about their procedures, their risks and benefits. Given this, it is not likely that they can glean knowledge that is anywhere near 'perfect', nor adequate enough, for them to make well-informed decisions about cosmetic surgery.

A risky business?

Risk and informed consent are key elements in the practice of cosmetic surgery, as in any other medical intervention. Risk is an inevitable part of any medical procedure. Whether we are undergoing a simple injection or major surgery, there is a risk that something may go wrong and that we may suffer some physical or psychological damage as a result. Patients who need to undergo a medical or surgical procedure in order to save their lives or to enable them to carry out normal daily tasks, weigh the possible benefits of their procedure against the possible risks. These risks may be outlined to patients by their medical team, as will the likelihood of success, given their health profile. At the same time, the medical team will be using their clinical judgements as to the efficacy of different treatments for individual patients. They will then seek the consent of the patient to follow a certain course of action.

When patients choose to access non-therapeutic procedures, such as cosmetic surgery, the element of risk is just as prevalent as for other medical procedures. Doctors carrying out cosmetic procedures are required to outline risks to their patients and to obtain their consent for those procedures. Risk can be related either to the surgical procedure per se, for instance risk linked to anaesthesia, or to the particular procedure that is being carried out. However, there are often differences in the way risk is conceptualised by patients when the procedure is cosmetic. Women in this study understood the usual risks of surgery and those associated with anaesthesia. However, the particular risks of certain cosmetic procedures were often not explained to them and, therefore, not understood by them. Unlike other patients, cosmetic surgery patients have diagnosed their own 'problem', they are prepared to invest a substantial amount of money to have their 'problem' fixed and enter the clinic with some idea of what they want from the surgery. At the same time, their entry into the medical process is often embarked upon with little information and with little understanding of what risks they are exposing themselves to. While these women may have been 'rational decision makers' (Fraser, 2003b: 87) when making the choice to have cosmetic surgery, the lack of information about the risks associated with that surgery meant that these decisions were too frequently made with little understanding about the possible adverse consequences of the surgery. Many women in this study were informed about procedures and practitioners through the media and publicity materials about cosmetic surgery. This material often characterises women as heroic, aware and fearless (Fraser, 2003b) agents who have an entrepreneurial approach towards their body and its possibilities. This approach is again reinforced through the utilisation of computer imaging. Here, women view their alternative bodies and visualise what they could have if they make the choice to have surgery. The doctor is the technician who can provide them with this body.

When things go wrong in cosmetic surgery, women are often characterised as victims. The women-as-victim characterisation is often found in magazine articles about cosmetic surgery failures (Fraser, 2003b). Some of these women are characterised as victims of their own vanity, while others are victims of poorly qualified practitioners. This categorisation did not sit easily with the women who shared their experiences in this study. Many presented themselves, not as victims of vanity, but as having been let down by practitioners who did not communicate well with them, did not outline the risks adequately, gave a technically poor result or gave an aesthetic result that was not wanted. As was discussed in Chapter 5, there is a tendency among cosmetic surgery practitioners to see adverse outcomes in cosmetic surgery as the responsibility of poorly qualified practitioners. They place the blame on those practitioners and on those women who select them. They argue that if women choose to use those practitioners, then they are ultimately accountable for this choice. There is little reflexivity in this approach and these practitioners fail to account for failures other than those perpetrated by practitioners outside their particular discipline.

The doctors interviewed for this study identified risk as, either something that was ever present in any surgical procedure or as directly related to the patient, their motivations and expectations. Those related to patient characteristics will be discussed later in this chapter. First of all, however, it may be useful to contextualise medical risk in the Australian setting and identify how women and doctors in this study reported how these risks were explained and understood.

Nisselle (1999) reports that the litigation rate in Australia involving medical mishaps increased by approximately 50 per cent between 1992–6. Data from the Medical Indemnity Industry Association of Australia (Pirani, 2005) show that legal claims against cosmetic surgeons rose from 233 during 1995–8 to 401 during 2001–4, an increase of 72 per cent. This was the largest percentage increase for any group of medical practitioners.

Nisselle also reports the result of a survey of patients and their relatives who were taking legal action. The survey found that there were four main reasons for the commencement of legal action. These were: deficiencies in the standard of care, a lack of information, need for accountability and seeking compensation for loss. Of particular interest is that a third of the recipients reported that they had litigated in order to obtain more information about their case. This group felt that information had only been given to them 'begrudgingly, incompletely and defensively' (Nisselle, 1999: 131).

The legal situation regarding the disclosure of risk by medical practitioners has been clarified in Australia through the case of *Rogers* v. *Whittaker* (1992). Nisselle (1999: 133) sees this case as a:

> watershed judgement that sign-posted the death of medical paternalism. It endorsed, as a matter of law, the patient's paramount right to know the material risks associated with medical treatment. In terms of medical practice, it expressed good sense and reflected contemporary 'best medical practice'.

The case centres on whether a treating doctor should have warned the patient of the small risk of blindness resulting from the procedure being carried out. As it transpired, the patient did lose her sight and argued that, had she been told of this risk, she would not have consented to the surgery.

It is the duty of medical practitioners to provide patients with adequate information about any material risk that a procedure may entail, and to let the patient decide whether they are prepared to accept the risk. There are a number of problems that are commonly cited that arise here. The first is related to the communication of risk. As outlined earlier in this chapter, there is now plenty of evidence to show that medical practitioners are too often poor communicators. Secondly, allowing time for patients to ask questions and seek clarification is often overlooked, as is the power imbalance between doctor and patient. In mainstream medicine, doctors often focus on diagnosis and treatment to the detriment of communication. However, here there may be therapeutic imperatives that justify such an approach. In cosmetic surgery, there is no therapeutic diagnosis that needs to be made and no health imperative that justifies the doctor making decisions on the patient's behalf, either about what information they should have or, about the nature and extent of their treatment. The patient makes the diagnosis before they enter the doctor's clinic. The doctor's role is to advise whether a procedure is possible and feasible, and to clearly and comprehensively outline the risks involved.

Risk and breast implants

One striking example of a failure to explain a known risk of a procedure is seen in the case of breast implants. There is a known risk of encapsulation with breast implants, where breast scar tissue can harden and move the implant. To avoid this, breasts need to be massaged regularly by the patient after surgery. Most of the women who had breast implants reported that they had not been advised about the risk of encapsulation before surgery, nor had they been told that they would need to massage their breasts. This is significant as it is a known and accepted risk. However, women reported that massaging the breasts was very painful. In many ways this example is indicative of an approach to risk discourse in cosmetic surgery that seems to underplay the risk to the body, and the pain, of undergoing these procedures in favour of emphasising the benefits that they can bring. The lack of information about breast implants and the risks they may pose is interesting given the history of silicone implants. One of the women interviewed had her silicone implants removed because of the problems she encountered with them. All other women reported having saline implants. However, it seems that the problematic history of silicone implants had little or no effect on the decision of women in this study to have implants.

Like inter-uterine devices, silicone breast implants were thought to be safe enough for many years until problems started to emerge (Rowsell et al., 2000). While medical professionals and their patients were not aware of these risks, information about the origin of silicone implants may have disturbed them. In a book that details the history of silicone breast implants produced by Dow Corning, John Byrne provides an interesting description of how silicone began to be used. Even though Dow Corning did not introduce its first breast implants until 1963, silicone may have found its way into the breasts of women as early as the late 1940s. In the aftermath of the Second World War, transformer coolant made of silicone was suddenly disappearing from the docks at Yokohama Harbour in Japan. The silicone fluid was used by cosmeticians to enlarge the small breasts of Asian prostitutes who knew that a more Western appearance would enhance their appeal to American service-men. Large doses of the doctored industrial fluids were injected directly into their breasts. To prevent silicone from migrating in the body, the Japanese added cottonseed or croton oil to cause immediate scarring, a way to contain the silicone at the site of the injection (Byrne, 1996).

The problems associated with silicone implants led to protracted legal proceedings in a number of countries, to significant public controversy and debate, and the filing for bankruptcy by Dow Corning, the manufacturer of the implants. During the debate on the safety of silicone implants, the stakeholders took varying positions. While lawyers, and some doctors, acting on behalf of women claiming to have suffered damage from implants argued that they were unsafe and toxic to the body, others argued that there were no independent scientific data to verify this claim. In particular those who manufactured the implants, and the doctors who implanted them, claimed that three decades of use proved that silicone implants were safe. At the same time, America's Food and Drug Administration, while acknowledging that anecdotal evidence was mounting against silicone implants, did not find sufficient scientific data to prove any toxic or other adverse effects. The debates as to the safety of the implants continue, and provide an insight into how different players interpret risk. Thus, while many scientists argue that there is insufficient evidence to link silicone implants to various diseases experienced by women who have or had them, others argue that there is sufficient cause for concern to limit their use. While there is little published data about the risks posed by saline implants, some expected risks may be: encapsulation, rupture, infection and change in sensation in the breast. Breast implants can also interfere with mammography. The women in this study were either not informed about such risks or were willing to accept those that they knew about.

The risk of an unsatisfactory aesthetic outcome

The medical risks emanating from surgery, or from particular cosmetic procedures, are not the only risks faced by women who elect to have cosmetic surgery. The risk of not getting the aesthetic outcome they want, or of looking and feeling worse than they did before surgery, is significant. The discourse of cosmetic surgery and the rhetoric surrounding its marketing allows little space for aesthetic failure. When such failure is discussed, it is either in the more sensationalist glossy magazines or in newspapers where it is linked to 'untrained' and 'unscrupulous' practitioners[1]. Again, doctors do not talk to patients of poor aesthetic outcomes. This is not part of the cosmetic surgery dialect. What they do discuss lies firmly within the area of patient expectations and responsibility. In discussing risk within cosmetic surgery, doctors see 'unrealistic expectations' and patient motivation as significant. The discourse here is about patient responsibility over their body. There is a significant disparity and tension here. On the one hand, doctors offer themselves to women as experts and artists who can transform the body to the individual's choosing. On the other, the responsibility for the choices made about that body lies with the woman and not with the doctor who carries out the procedures. While this maintains what Fraser describes as the 'repertoire of agency' (Fraser, 2003a; 2003b) which permeates popular, medical and some feminist (Davis, 1995) literature on cosmetic surgery, it shifts responsibility from the doctors for the outcomes of such surgery and places it firmly with the patient. This is where doctors can rely on explaining away poor aesthetic outcomes on the basis of unrealistic expectations, or as being acceptable, because the result was technically satisfactory. Fraser (2003b: 172) identifies the trend to place accountability in cosmetic surgery with the patient in a range of medical materials on cosmetic surgery. She says that these 'repertoires':

> function more or less explicitly to render the individual responsible in relation to practices that are controversial in terms of safety. At the same time, they often insert the authority of the surgeon or other medical practitioner into the process of decision-making while leaving ultimate accountability with the consumer.

This is also linked with what motivates women to have cosmetic surgery. Doctors were adamant that women should have cosmetic surgery for themselves, and not because they were under pressure from anyone else. Women in this study and others (Davis, 1995), emphasised that they wanted cosmetic surgery for themselves and not to please anyone else. Internal motivations are again applauded by those feminists (Davis, 1993; 1995; 1998) who reiterate the agency in the act of body modification. One doctor took this further by stating that good results and bad results were not really the issue in cosmetic surgery: what was relevant were 'happy and unhappy patients' (Dr Inch). This suggests that the responsibility for the outcome of cosmetic surgery lies not with the doctor but in the fundamental character of the patient. This will be discussed more fully later in this chapter.

Women, however, saw the matter of an unsatisfactory aesthetic outcome very differently from the doctors. The most striking examples of this were the women that had breast implants. Not only was the medical risk associated with implants poorly explained, but the fact that implants were larger than women wanted was significant. Of most importance to women was the fact that instead of making them look and feel as if they fitted in to normative standards of the 'ideal' breast size, they now stood out because their breasts were too large. While some of these women were not totally happy with their new breasts, they felt they were better than what they had before and they balanced the outcome of their surgery with the overall process they had been through. If they were satisfied with the relationships and communication they had with their doctors, then they accepted the outcome they had been given. If, however, their relationships and communication with their doctors were not good, then their dissatisfaction with the outcomes of their surgery was significant.

Another theme that was common in the narratives of these women was that they took responsibility for the choices they made and were prepared to accept the consequences of those choices. What they could not accept was the imposition of an aesthetic outcome on their bodies that was not of their choosing, when it was coupled with the doctor's inability, or refusal, to communicate with them meaningfully and to treat them with respect.

This was again reinforced by women who had been offered, or who had undergone, additional cosmetic procedures. Of course, there is a fine line here between what the doctor, from his experience, believes will give women the outcomes they are seeking and encouraging women to have additional procedures that go beyond what they really want. Here the 'objectifying gaze' (Bartky, 1998; Foucault, 1973; Hillier, 1997) of the doctor creates the 'docile' body (Foucault, 1979; Morgan, 1991), where the doctor's vision of aesthetic acceptability is imposed through suggestion to the woman. This again erodes a woman's agency over her own body and creates a tension between what she wants for that body and what the doctor suggests she should opt for. It is paradoxical that, as we are able to choose more forms of cosmetic intervention to change and control our bodies, we are also subjecting our bodies more and more to the control of science and medicine. As Shilling (1994: 4) writes:

> As science facilitates greater degrees of intervention into the body, it destabilises our knowledge of what bodies are and runs ahead of our ability to make moral judgements about how far science should be allowed to reconstruct the body.

While medical science continues to develop new ways to change women's bodies through surgical and technological intervention, those changes are implemented by doctors whose views about women's bodies, and the visions they have of them, are culturally influenced and socially delineated. This, in turn, leads to outcomes that are not wholly determined by the wishes of women, but also influenced by the attitude of the doctors. This study has shown how significant the doctors' influence is on the aesthetic outcomes women achieve from cosmetic surgery.

Doctors' and women's perceptions of risk in cosmetic surgery

It is clear that there are significant differences in the way women and doctors conceptualised risk in cosmetic surgery. Doctors identified risk as either something that was ever present in any surgical procedure or as directly related to the patient, their motivations and expectations. Research carried out during the 1970s and 1980s found that professionals and the lay population assess risk differently (Gabe, 1995). Psychological research on risk has been criticised for focusing too narrowly on individual approaches and attitudes and not allowing for social or cultural factors and influences. Gabe (ibid.) reports on research carried out on couples undergoing fertility treatment. This research shows how perceptions of risk are socially and culturally influenced. Thus, the women's desire to have children (both socially and culturally created and reinforced) overwhelmed perceptions of any associated risk. Similarly, it can be argued that women who choose to undergo cosmetic surgery do so within the context of the socially constructed ideas of how they should look, or how they *feel* they should look in order to fit comfortably into society. Thus, while the medical practitioner might explain to them certain *technical* risks associated with the procedure they will undergo, for them, these risks do not rate as highly as the risk of doing nothing and continuing to look and feel the way they do. Some women said that they would have gone through with the procedure despite the risks involved. Others stated clearly that, if they had known all the possible risks, they would not have continued with the surgery.

Women saw risk from a broader perspective than that which is usually assumed in the medical context. Thus, the risks of anaesthesia, the risk of an adverse technical outcome or unexpected medical complications were all cause of concern, but they were aware of these things before they entered the clinic. Also a cause of concern, however, was the recovery process, the pain and discomfort resulting from a procedure, risks associated with breast implants and the risk that the result might not be what was expected. Many felt that these risks were not adequately communicated to them. Other risks were either not identified or not clearly delineated. In many ways it is not surprising that doctors do not provide information on pain and similar possible consequences of a cosmetic procedure. They are caught in a conundrum where cosmetic surgery is, on the one hand, identified as real surgery that should only be carried out by appropriately qualified practitioners, but on the other is marketed as accessible, straightforward and a realistic option for any woman who chooses to improve her body. Doctors identified themselves as artistic and creative, as sculptors of the female form. Again, this identification does not sit well with images of pain and scarring, or of a body oozing blood and other fluids.

We have already seen that doctors emphasised the risk of unrealistic expectations among their patients as significant in cosmetic surgery. But, it was not really clear from the narratives of the doctors what they really meant by this. Dr Green mentioned that he limited patients' expectations by saying that he could not 'turn Lassie into Marilyn Monroe'. This comparison of a dog to a film icon seems flippant and somewhat insensitive to say the least. How are women to interpret such a statement and in what way does it explain what is possible from cosmetic surgery? With information such as this can women actually make a well considered choice about what cosmetic surgery can do *for* them and what they allow the doctor to do *to* them?

Issues relating to informed consent

In all medical procedures informed consent has practical and ethical connotations. In practical terms, patients need to understand the medical treatment being proposed, what it entails and its likely consequences. Doctors need to be sure that the patient understands what is going to be done to them so that they cooperate with the course of treatment. Consent does not appear as a doctrine in classical medical ethics, but has developed in response to notions of the rights of individuals to have a say about what is done to them and their bodies (Veatch, 1999).

Traditionally, and certainly before the twentieth century, there was no notion of patient input into medical decision-making (Veatch, 1999). There is no reference to consent in the writings of Hippocrates. However, the right of the patient to understand and agree to treatment is now accepted and legally required. In order to give consent, patients need appropriate information about their treatment. How much information is enough is, however, often contentious. McClean (1989: 8) sees 'adequate information disclosure' as a 'moral' rather than 'purely technical' issue. Thus:

> If information disclosure is genuinely to protect patient autonomy, then obviously all relevant information should be disclosed. Thus, every known risk – whatever its statistical probability – should be made known to patients.

For McClean, patient autonomy and protecting that autonomy are paramount. Traditionally, medical practitioners have tended only to provide patients with information which they feel is relevant, and they have attracted criticism for being too 'paternalistic' because of this. However, selective provision of information may be inadequate to enable patients to provide truly informed consent. McClean argues that in order to protect the rights of patients and enable them to provide informed consent, information disclosure on two levels should be required. These are disclosure of therapeutic alternatives and disclosure of risks and benefits.

Patients would be informed of available choices and evaluate what is best for them. As a result:

> It is not inconceivable that a patient may be prepared to accept the risks associated with one therapeutic option but not with another.
>
> (McClean, 1989: 8)

Somewhat less important risks may be overlooked by the doctor, but may be significant to the patient because it is they who have to suffer the consequences. For instance, the risk of significant bruising, slower than expected recovery and healing may not be regarded as significant to the doctor, but may be significant to the patient. It may well be that, if told of these risks, a patient may give significant weight to them. In this study, many women identified risks such as these as worthy of mention by a doctor despite the fact that they had not been mentioned to them by their doctor.

Although there is a legal and an ethical requirement to obtain informed consent, the means through which this is achieved is not mandated and there are no clear, consistent, standardised or comprehensive legal guidelines to follow. The key elements of informed consent are competence, voluntariness, disclosure of information and understanding. Relevant Australian judgements do provide guidance about the means by which doctors provide information about risk to patients and obtain consent. Skene and Smallwood (2002:40) note that

> [a] doctor cannot discharge the duty to inform simply by providing pamphlets about a proposed procedure, such as a pamphlet mentioning 'capsulation', infection, asymmetry, or change in nipple-breast sensation as risks of breast enlargement, or the Australian Society of Plastic Surgeons' brochure, Patient Information on Abdominoplasty … taken from the doctor's surgery.

Some of the doctors in this study did rely on such pamphlets to provide information to their patients and, as such, it is doubtful whether they would meet the legal test for achieving informed consent. Doctors also mentioned using practice staff to gain consent from patients. This distances the doctor from the actual act of obtaining consent and makes it a mere technical formality rather than an ethical and legal agreement reached between doctor and patient. Practice staff cannot be expected to understand all the risks involved in surgical procedures, are not liable for the consequences of those procedures, and have a limited ability to provide enough information to elicit informed consent.

Davis (1995: 147) reports that women in her study were unhappy about the circumstances in which they made decisions about their surgery and:

> [r]egardless of the outcome of the surgery, many women indicated that they were angry at not having been consulted about what they wanted from the operation.

Shades of medical paternalism and the failure of doctors to actually detail the outcomes of surgery and its risks were evident in Davis' research, as they are in this study. Lack of adequate knowledge before surgery and a lack of information provision about the surgery were also evident in women's narratives. Such an information gap limits the possibility of giving informed consent. Davis (ibid.: 157) maintains that

> The conditions for informed consent are rarely adequate. Most women decide to have cosmetic surgery without being able to discuss their problems and their options with others who are sympathetic to their situation. They do not have the opportunity to make the optimal assessment of the operation and its drawbacks. This is not only because the available information is inconclusive or because there is a certain amount of risk involved in surgery. Physicians systematically withhold information or downplay the risks of surgery.

Davis' findings are remarkably consistent with those of this study, where women either reported that certain risks had not been explained to them, or that risks were downplayed by doctors and their staff.

Many of the narratives of the women suggest that the information given to them about their surgery and its risks, and the way that information was given to them, was not comprehensive and not adequate to ensure their fully informed consent. For instance, when reviewing the medical and scientific responses to the issue of the safety of breast implants, Fraser (2003b: 127–8) argues that

> In a clearly paternalistic mode, surgeons position women as in need of information only up to the point that they feel reassured about the procedure's safety. This suggests that a decision about safety has been made prior to women's involvement and that participants are actually informed about the outcome of that decision, which is that the procedure is safe. Informed consent, posed by many as the appropriate response to risky operations, means little in this context.

This study has outlined the variety of ways doctors sought informed consent and what information women were provided with before consenting to a procedure. Many women reported significant gaps in the information provided to them. This implies that many women underwent cosmetic surgery with incomplete knowledge about risk and information that was not adequate enough to ensure truly informed consent.

Women reported that they often did not fully appreciate all the risks involved in cosmetic surgery, many significant risks were not explained to them and, when risks were explained, the means through which this was done was unsatisfactory. Doctors reported that they did explain risks to women, but, in common with the women's narratives, they said that they often used practice staff to undertake this task. If risk is not adequately or comprehensively explained to women, then they are unable to make fully informed decisions about their body, or the procedure that they are about to undergo.

Choice and limitations in cosmetic surgery

At the heart of all aspects of the cosmetic surgery process is the issue of choice. Some feminist theorists (Davis, 1993; 1995; 1997b) maintain that, as autonomous individuals, women should be allowed to make choices about their bodies and exercise agency in making those choices. Others (Bordo, 1993; Dull and West, 1991; Morgan, 1991; Wolf, 1990) contend that women's choices about cosmetic surgery are made against a background of objectification of women's bodies and a cultural imperative that allows little variation in female beauty. Fraser (2001; 2003a; 2003b) reflects these debates in her analysis of a range of popular, academic and medical publications and of the advertising materials that promote cosmetic surgery. While women's abilities to make choices about their bodies is fundamental to their decision to have cosmetic surgery, and it is acknowledged that those choices are made within particular cultural and social contexts, the limitations of those choices have not been adequately explored. That is, women's decisions about their bodies are subsumed in a 'Pandora's box' (Sullivan, 2001) of issues that render those choices limited and open to the myriad of influences that infuse the practice of cosmetic surgery.

It was clear that the choices women made about their bodies were for themselves rather than for the gratification of another. Despite the autonomous nature of such choice a myriad of influences came into play when they engaged with the doctor in the process of cosmetic surgery, and these influences have been outlined in Chapters 7, 8 and 9. What we see here is that women's choices over their own bodies were often manipulated and affected by the very process of cosmetic surgery, and by the interactions that occurred within it. There is a long history of women's bodies, and in particular their fertility, being controlled by medical intervention. Cosmetic surgery provides yet another example of such 'colonisation' (Morgan, 1991).

Cosmetic surgery comprises a complex web of interactions and decisions. Many women were prepared to accept the consequences of their choices and bear responsibility for their actions, whatever the outcome. Others despaired of what had happened to them and felt that they were now worse off than they had been before their surgery. They admonished themselves for believing the rhetoric of cosmetic surgery and for believing doctors who let them down.

Doctors who choose to practise cosmetic surgery are engaged in a lucrative and commercially driven enterprise. Doctors explained their choices to do cosmetic surgery as both internal, driven by personality and artistic temperament, and external, offering a psychological and medical service to women. However, they shied away from engaging with the commercial reality of what they did. Women, on the other hand, were cognisant of the economic rewards of cosmetic surgery and saw this as significant motivating factor for doctors. Doctors were more interested in debating who should be carrying out cosmetic surgery rather than reflecting on their role in an industry which is often seen as preying on women's vulnerabilities about their bodies. Nevertheless many doctors in this study characterised themselves as 'altruistic healers' (Sullivan, 2001: 195) whose role as cosmetic surgeons provided a psychological, as well as physical, benefit to women.

There is an 'institutionalized schism' (Sullivan, 2001: 56) that pervades the industry and that engages in debate about who should rightly be carrying out cosmetic surgery. The consequence of this is that rather than participating in significant and systematic steps to make the industry more accountable, consumers better informed and the process of cosmetic surgery more transparent, practitioners gaze inwardly at each other and pick apart those that they see as not belonging to their particular group. This limits their ability to provide a service that is in the best interests of their patients and, ultimately, the industry in which they practise.

Women's choices are also limited by social and cultural prerequisites and by the way these are interpreted through the eyes of the surgeon. There are numerous examples here of how differently women and doctors explain their experiences of cosmetic surgery. It was often as if each group were engaged in different, rather than shared, experiences and in disparate rather than common discourses. Keeping in mind the historical power of medical authority over the body, it is hardly surprising that the degree of autonomy women can maintain in cosmetic surgery is only as significant as that authority allows it to be. Fraser (2003b: 193) concludes that

> [f]rom this point of view, women's 'autonomy' comes to be rearticulated within the supposedly reasonable limits set by the surgeon. This is an interesting dynamic because while femininity may be

being redrawn with greater autonomy, the 'bottom line' reiterated again and again is male authority. Through cosmetic surgery as a technology of gender, very conventional gender dynamics are played out within an apparently liberal script of women's self-definition.

The method through which the surgeon establishes and maintains authority is mainly through language. The discourse adopted and used by the surgeon is based in, and reflective of, the discipline of medicine. Foucault (1972: 224) claims that 'Disciplines constitute a system of control in the production of discourse'. He also maintains that particular discourses, such as 'therapeutic' (medical) ones 'are ... barely dissociable from the functioning of a ritual that determines the individual properties and agreed roles of the speaker' (ibid.: 225). In cosmetic surgery the 'individual properties', the fact that most practitioners are male, and the 'agreed roles', their roles in changing women's bodies through surgery, create a discourse that is powerful and authoritative and which translates the wishes of women concerning their bodies, both through male eyes and through a medical paradigm.

In cosmetic surgery women's bodies provide a canvas for doctors to engage in the art of surgery and to make a handsome profit at the same time.

Frustratingly, they seem to lack serious or systematic reflection about their trade, or about the processes and actions required to provide quality and dispassionate information to patients. The deliberations about who should carry out cosmetic surgery detract from the real issues of whether doctors communicate well enough with their patients, provide enough information about risk and allow patients to be equal participants in decision making about their bodies. While doctors and women engage in such distinct and different discursive practices, there should be continuing concern about the efficacy of cosmetic surgery as a solution for women who feel ill at ease with their body.

Conclusion

Cosmetic surgery is one of the solutions women seek to be at ease with who they are and how they look. Why some women seek surgical intervention while others don't is not clear, but is probably linked to a complex set of reasons about how we all feel about ourselves and our place in society. Of course, cosmetic surgery is essentially gendered and any change to this would require fundamental changes in the social and cultural position of women. The choices women make about their bodies have to be contextualised in relation to the cultural imperatives and pressures they face. The reasons why they make such choices have to do with the way they feel, as individuals, about their body and the way they fit into the world around them. It also has to do with the unrepenting surveillance that the female body is subjected to, both overtly and covertly. This surveillance is reflected through the surgeon's scalpel where it is translated into the bodily reality that creates a female body that is more acceptable to current standards of female beauty. A variety of bodily customs and modifications have been a constant reality for women across history, but cosmetic surgery has engaged the services of the medical profession and made such modifications a medical matter.

Doctors use the discourses of medicine and medical diagnosis as tools to reconceptualise what is, in reality, a dissatisfaction with, rather than disease of, the body. As doctors are trained to deal with disease and illness it is not surprising that they translate women's feeling about their bodies into medical talk that justifies and contextualises what they do as medically based. It is these medical discourses, and the gendered nature of understandings of women's bodies, that often result in confusion and incongruent communication between women and doctors. The selectivity of the risks that women reported were explained to them is indicative of much that was unarticulated by doctors in the cosmetic surgery consultation. Despite what doctors reported, many women felt ill informed and dissatisfied on a number of levels.

In the end, the objective measure of the success of any cosmetic surgery procedure is directly related to patient satisfaction with the outcome and the process of surgery. The best that can be said of what so many women reported was that they were just 'satisfied enough'. Is this really satisfactory for an industry that is becoming pervasive in Western countries and blooming in developing countries that are able to provide surgery at less cost? Shouldn't we becoming concerned about the potential the industry has to maim whilst promising psychological benefit?

Despite the fact that advertisements and popular media, and sometimes doctors, might report an accessible and straightforward journey from dissatisfaction with one's body to satisfaction, the reality can be disturbingly different.

What is clearly articulated in this study is that the rhetoric surrounding cosmetic surgery often belies its complexity, and the intricate nature of the engagement between the doctor, the clinic and the patient. There are many contradictions in the narratives of women and doctors about cosmetic surgery and their engagement in the process. These contradictions expose the limitations of cosmetic surgery as a tool through which women can achieve an outcome that *they* want from that surgery. It is a process where women are both accommodated and disciplined (Fraser, 2003a) where they are provided with a solution to their dissatisfaction with their body, so long as it is within the boundaries prescribed by medical authority. This is not clearly understood by women, nor articulated by doctors. Instead both parties set off on a course of action that is often fraught with misunderstanding, and whose outcome is dependent on a plethora of influences that are complex and disparate. The result of all this is a bodily reality for those women who choose cosmetic surgery. For some the gamble is worth it, but for others it ends in disappointment and further alienation from their bodies.

10
Looking Forward ... Looking Back

What we have in cosmetic surgery is a conundrum. If cosmetic surgery is decried as an unnatural practice that subjugates women by making them mere vessels of beauty imperatives imposed by the media and society as a whole, we deny women the intelligence and decision-making faculty to make choices about their body. If, on the other hand, we recognise it as an acceptable practice and one that rightly enjoys the time and attention given to it by medical practitioners, then we are supporting the right of doctors to interfere in yet another facet of women's lives, to make substantial amounts of money from this and without too much scrutiny of, or medical justification for, what they are doing. Neither position is tenable so we sit in an uncomfortable space between the two when we try to face the complex questions that cosmetic surgery poses. Theorists have too often drawn a demarcation line between social and cultural pressures on women and the right of women to make choices about their bodies. They have argued among themselves rather than turning their critical gaze on the doctors whose business it is to change women's bodies. What we find in looking at the process of cosmetic surgery is that, whether seen as agents or victims, their initial motivations for deciding on cosmetic surgery is only the beginning of a complex journey for women to an outcome that they may, or may not, want or accept. This outcome is in the hands of the doctor and the uncertainty of cosmetic surgery for women is a significant but silent factor in the process.

Various texts including popular media, health media and medical texts discuss cosmetic surgery (Fraser, 2001; 2003a; 2003b). Various notions such as agency, vanity, artistry and the 'natural' are delineated through these texts. Many of these are mirrored in women's interactions with doctors. Thus, Fraser (2003b: 193), in her analysis of 'agency repertoires', maintains that

> these versions of agency almost without exception position the surgeon in a paternal role, providing advice, approving or veto-ing aesthetic choices, invested with the power to withhold surgery altogether ... the 'bottom line' reiterated again and again is male authority. Through cosmetic surgery as a technology of gender, very conventional gender dynamics are played out within an apparently liberal script of women's self-definition.

The narratives of women and doctors described in this book have demonstrated that the doctor's role in the decisions about surgery, on whether to operate and on the outcomes of surgery, is significant. As we have seen, the process of surgery is made up of a multifaceted con-struct that includes the interactions women have with doctors, their staff and the technological tools of cosmetic surgery. From the medical perspective, these mechanisms are all there to enhance and explain the cosmetic surgery experience and to enable it to be a seamless transition to an improved body. However, the complexity of human interaction, of power differentials and of understandings and misunderstandings about cosmetic surgery influences how women experience and interpret them and how they make sense of them.

There is no one aspect of the cosmetic surgery process that can be changed to make the outcome more accepted by, and acceptable to, patients and doctors. Cosmetic surgery is fundamentally involved with, and reflective of, gender constructs in our culture and, while patients are predominately women and doctors predominately men, those con-structs will be replayed in cosmetic surgery clinics on a daily basis. Even when the practitioner is a woman, there is little guarantee that she will understand what women want. Here, female practitioners work very much within the male paradigm that dominates specialist medicine and medical intervention with women's bodies has a long and chequered history. What is new in cosmetic surgery is that this medical interven-tion is the site at which the demands of commerce, the services of medi-cine and the cultural demands on women's bodies meet.

Cosmetic surgery is first and foremost a commercial endeavour. The history of cosmetic surgery shows the links it has with the need people have to 'pass' in society (Gilman, 1998; 1999). A major impetus for the widespread availability of cosmetic surgery came with the changes in regulations governing advertising for medical practitioners. Sullivan (2001: 189) says that many practitioners were 'seduced by the power of commercialism, unleashed by deregulation'. At the same time, the selling of cosmetic surgery as a commodity both responds to and acts to generate demand. In this way it is not unlike the advertising and marketing of any other consumer product. Doctors in this study explained how they marketed their services, and while some depended on 'word of mouth' for referrals, many advertised their practice in a range of other ways. In fact, some felt that if they did not advertise, they would lose business to those who did. This meeting of medicine and the market is often ethically problematic for practitioners. In stimulating demand and creating the expectation among patients that they can change their bodies through a straightforward and seamless process, they mask the reality of invasive surgery which is often painful, difficult and which can have unexpected outcomes.

Many practitioners argued that the risks of cosmetic surgery were made clear when the patient came to the clinic. However, we have seen that women's recollections of the explanations of risk were often not in tune with the reality of the risks they faced or experienced. It is not clear whether many of the women in this study would have gone ahead with cosmetic surgery if they had known and understood the nature and extent of the medical risks they faced. The risk of not actually achieving the aesthetic outcome they wanted from cosmetic surgery, of course, is something women tend not to focus on and is invariably glossed over in the clinic. While a range of technological tools and more traditional methods may be used to reach an agreement about the aesthetic outcome of surgery, in the end, it is the doctor who carries out the procedure and is the final arbiter of what is achieved. The outcome is influenced by both the technical skill of the doctor and by his vision of what is acceptable. While many women were satisfied enough with their outcome, many were not because they felt that what they ended up with was something that the doctor gave them, rather than what they thought they had negotiated.

The complexity of negotiating the cosmetic surgery process depends on the communication between doctor and patient. The varied approaches to communication that exist in this setting often act as roadblocks to effective communication between women and their doctors. Given what we already know about doctor–patient communication, it is of little surprise that in cosmetic surgery, as in many other areas, misunderstanding and failure are at the heart of dissatisfaction with this medical practice. It is significant that if women felt that they have good communication with their doctors, they were prepared to be forgiving of outcomes with which they were not particularly satisfied.

While many of the doctors interviewed mentioned that good rapport with patients is vital, it is not clear whether their conceptualisation of what this means is the same as that of their patients. There are many examples that suggest that doctors and patients failed to understand each other, where shared understanding is not achieved. This has to do with the gendered nature of discourse and with the influence of medical power and authority on the interchange, and is embedded in the very nature of cosmetic surgery. In the end, sharing a vision of how we want our bodies to look with another, and asking that person to change our bodies to our exact requirements, is complicated and difficult to achieve. If practitioners were more open and honest about their limitations and fallibility in this area, then patients might be more understanding when things do go wrong. However, this is not something that is likely to be widespread while cosmetic surgery remains a commercially driven and promise fulfilling endeavour.

As in any profession, there are some cosmetic surgery practitioners who are better than others at what they do. Some of this has to do with technical skill, and some to do with the ability to communicate meaningfully and effectively with patients. While the cosmetic surgery industry continues to focus on fighting a 'turf war' between plastic and reconstructive surgeons and all the other practitioners who practise cosmetic surgery, the industry is not paying attention to the issues that concern patients. That is, they are not engaged in reflecting on how they can provide better information, communicate more effectively and offer a more comprehensive understanding of the risks of procedures to patients. Nor are they really focused on why patients want cosmetic surgery. This project has shown that all these factors are important to patients and are integral to successful outcomes in cosmetic surgery.

It is astonishing that cosmetic surgery has previously attracted so little empirical investigation. While this book has sought to contribute to understanding the experiences of women and doctors in the process of cosmetic surgery and has exposed some significant discrepancies in their understanding of each other, it has also exposed how little we know about the cosmetic surgery industry. Much of the latter seems 'hidden' within clinics and away from any significant scrutiny and many women who undergo cosmetic procedures do so in relative secrecy and with little information. The significant role that practice staff play in the industry became clear from the reports of both women and doctors; but nothing is known about the details of their work. For the first time, women's and doctors' reports of their experiences of cosmetic surgery have been intricately detailed in this study. However, we have no information about how they *actually* interact in the consultation. Nor do we know the real impact of technologies, such as computer imaging, on women's decision-making process in cosmetic surgery. The list of what is not known about cosmetic surgery is extensive. Negotiating the 'normal' body through cosmetic surgery is an intimate and complicated process where the actors have little understanding of each other and of how each experiences the process of cosmetic surgery.

There is no one solution that can address, nor account for, these disparities. Nevertheless, there is a need for the cosmetic surgery industry to be more aware of the contexts and meanings in women's lived experiences that lead to their engagement with that industry. At the same time, women need far more comprehensive information about cosmetic surgery and, in particular, its risks. Such information will enable them to make better-informed choices about their bodies.

Better communication between doctors and women and a shared understanding of risk would improve the outcomes of cosmetic surgery. However, cosmetic surgery is a wholly new paradigm for the practice of the medical arts. Other medical specialities usually operate in a seller's market. That is, there are more patients needing care than doctors to meet their needs. For a non-therapeutic area like cosmetic surgery it is a buyer's market where practitioners actively seek patients to treat. Their work is not predicated on reducing the incidence of disease or on improving public health. Instead, it is based on the anxiety that patients feel about their bodies. We have seen that cosmetic practitioners perceive their work as psychologically therapeutic. However, we have no systematic evidence that this is, indeed, the case. There needs to be a new multi-disciplinary analysis of the cosmetic surgery industry that takes us beyond current ways of thinking and enables us to critically appraise what is actually happening within it so that such claims can be properly scrutinised and tested.

Notes

1. Introduction

1. I have included liposuction as a surgical rather than a medical procedure as it requires the insertion of a cannula into the fatty deposit beneath the skin and is carried out under either local or general anaesthetic.

2. Cosmetic Surgery and the 'Normalisation' of the Body: A Short History

1. See www.plasticsurgery.com.au/history.
2. The French cosmetic surgery texts of the time utilised such language to describe the aims of cosmetic interventions; see Comiskey (2004).
3. Described by B. W. Taylor, in 'The History of Australian Plastic Surgery', Australian Society of Plastic Surgery, www.plasticsurgery.com.au/history.

4. Social and Feminist Theories and the Body

1. See, for instance, Sullivan (1997), 'Domination and Dialogue in Merleau-Ponty's *Phenomenology of Perception*', Hypatia 12(1): 1–19.
2. See Eagleton (1993), It is not quite true that I have a body, and not quite true that I have one either (Review of Body Work, by P Brooks). *London Review of Books*. 27(5), 7–8.

5. Into Battle for the Cosmetic Surgery Market

1. Brighton Plastic Surgery Centre, Mr. Keith L. Mutimer F.R.A.C.S (PLAST). About Your Surgeon. (original emphasis).

9. Overview

1. See, for instances, 'Woman Wins Breast Case', *The Age*, 11 June 2005 and 'You too Could Look Like This', *The Sunday Age*, 3 July 2005.

Bibliography

Adams, A. E. (1997) 'Molding Women's Bodies: The Surgeon as Sculptor'. In D. S. Wilson and C. M. Laennec (Eds) *Bodily Discoursions: Gender, Representations, Technologies*. NY, State University of New York Press.

American Society of Aesthetic Plastic Surgeons (2003) Press Release.

American Society of Plastic Surgeons (2007) 2006 Gender Quick Facts Cosmetic Plastic Surgery. American Society of Plastic Surgeons.

Andriole, D. A., Klingensmith, M. E. and Jeffe, D. B. (2006) Who are Our Future Surgeons? Characteristics of Medical School Graduates Planning Surgical Careers: Analysis of the 1997 to 2004 Association of American Medical Colleges' Graduation Questionnaire National Database. *Journal of the American College of Surgeons*, 203, 177–185.

Arras, J. D. and Steinbock, B. (Eds) (1999) *Ethical Issues in Modern Medicine*. Mountain View, CA, Mayfield Publishing Company.

Arthurs, J. and Grimshaw, J. (Eds) (1999) *Women's Bodies: Discipline and Transgression*. London, Cassell.

Askegaard, S., Gertsen, M. C. and Langer, R. (2002) The Body Consumed: Reflexivity and Cosmetic Surgery. *Psychology and Marketing*, 19, 793–812.

Atkinson, P. (1995) *Medical Talk and Medical Work: The Liturgy of the Clinic*, London, Sage.

Australian Centre for Effective Healthcare University of Sydney (1999) A Review of Published Literature on the Effectiveness of Selected Cosmetic Surgery Procedures. Sydney, Australian Centre for Effective Healthcare.

Australian Institute of Health and Welfare (2003) A Growing Problem: Trends and Patterns in Overweight and Obesity 1980 to 2001. Canberra, Commonwealth Government.

Australian Medical Association (2004) Position Statement: Advertising and Public Endorsement – 2004. Editorially Revised 2006. Canberra, AMA.

Australian Medical Association (2006) Advertising and Public Endorsement Position Statement. Canberra, AMA.

Baker, N. C. (1984) *The Beauty Trap: Exploring Woman's Great Obsession*. NY, Franklin Watts.

Balsamo, A. (1996) *Technologies of the Gendered Body: Reading Cyborg Women*. Durham, Duke University Press.

Bartky, S. L. (1990) *Femininity and Domination: Studies in the Phenomenology of Oppression*. NY, Routledge.

Bartky, S. L. (1998) Foucault, Femininity, and the Modernization of Patriarchal Power. In R. Weitz (Ed.) *The Politics of Womens' Bodies: Sexuality, Appearance and Behaviour*. NY, Oxford University Press.

Becker, A. E., Burwell, R. A., Gilman, S. E., Herzog, D. B. and Hamburg, P. (2002) Eating Behaviours and Attitudes following Prolonged Exposure to Television among Ethnic Fijian Adolescent Girls. *British Journal of Psychiatry*, 180, 509–14.

Bigwood, C. (1991) Renaturalizing the Body (With a Little Help from Merleau-Ponty). *Feminism and the Body, Hypatia Special Issue*, 6.

Bishop, R. (2001) The Pursuit of Perfection: A Narrative Analysis of How Women's magazines Cover Eating Disorders. *The Howard Journal of Communication*, 12, 221–240.

Blake, C. F. (2000) Foot-Binding in Neo-Confucian China and the Appropriation of Female Labour. In L. Schiebinger (Ed.) *Feminism and the Body*. NY, Oxford University Press.

Bloor, M., and Avebury, P. T. (Ed.) (1994) *Qualitative Studies in Health and Medicine*. Aldershot, Avebury.

Blum, V. L. (2003) *Flesh Wounds: The Culture of Cosmetic Surgery*. Berkeley, University of California Press.

Bordo, S. (1993) *Unbearable Weight: Feminism, Western Culture and the Body*. Berkeley, University of California Press.

Bordo, S. (1998) Braveheart, Babe and the Contemporary Body. In E. Parens (Ed.) *Enhancing Human Traits*. Washington DC, Georgetown University Press.

Bradbury, E. (1994) The Psychology of Aesthetic Plastic Surgery. *Aesthetic Plastic Surgery*, 18, 301–5.

Braun, J. M. (2005) Just a Nip and a Tuck? The Culture of Cosmetic Surgery. *Feminism and Psychology*, 15, 345–50.

Brooks, A. (2004) 'Under the Knife and Proud of It'. An Analysis of the Normalization of Cosmetic Surgery. *Critical Sociology*, 30, 207–39.

Brownmiller, S. (1985) *Femininity*. NY, Fawcett Columbine.

Brush, P. (1998) Metaphors of Inscription: Discipline, Plasticity and the Rhetoric of Choice. *Feminist Review*, 58, 22–48.

Butler, J. (1993) *Bodies that Matter: On the DiscursiveLlimits of 'Sex'*. NY, Routledge.

Butler, J. (1997) *The Psychic Life of Power*. Stanford, CA, Stanford University Press.

Byrne, J. A. (1996) *Informed Consent*. NY, McGraw-Hill.

Bytheway, B. and Johnson, J. (1998) The Sight of Age. In S. Nettleton and J. Watson (Eds) *The Body in Everday life*. London, Routledge.

Cahill, H. A. (2000) Male Appropriation and Medicalization of Childbirth: An Historical Analysis. *Journal of Advanced Nursing*, 33, 334–42.

Carter S (1995) Boundaries of danger and uncertainty: an analysis of the technological culture of risk assessment. In J, G. (Ed.) *Medicine, Health and Risk*. Oxford, Blackwell Publishers.

Cash, T. F. (1990) The Psychology of Physical Appearance: Aesthetics, Attributes, and Images. In T. F. Cash and T. Pruzinsky (Eds) *Body Images: Development, Deviance and Change*. NY, The Guildford Press.

Cash, T. F., Ancis, J. R. and Strachen, M. D. (1997) Gender Attitudes, Feminist Identity, and Body Images Among College Women. *Sex Roles*, 36, 433–47.

Cash, T. F. and Deagle, E. A. (1997) The Nature and Extent of Body-Image Disturbances in Anorexia Nervosa: A Meta-Analysis. *International Journal of Eating Disorders*, 22, 107–26.

Cash, T. F., Duel, L. A. and Perkins, L. L. (2002) Women's Psychosocial Outcomes of Breast Augmentation with Silicone Gel-Filled Implants: A 2–year Prospective Study. *Plastic and Reconstructive Surgery*, 109, 2112–2121.

Cash, T. F. and Pruzinsky, T. (Eds) (1990) *Body Images: Development, Deviance and Change*. NY, The Guildford Press.

Cash, T. F. and Pruzinsky, T. (2002) *Body Image: A Handbook of Theory, Research, and Clinical Practice*. NY, The Gulidford Press.

Cassell, J. (1991) *Expected Miracles: Surgeons at Work*. Philadelphia, Temple University Press.

Castanares, S. (1977) Ethics in Aesthetic Surgery. *Aesthetic Plastic Surgery*, 1, 209–12.

Castle, D., Honigman, R. J. and Phillips, K. A. (2002) Does Cosmetic Surgery Improve Psychosocial Wellbeing? *Medical Journal of Australia*, 176, 601–04.

Chait, L. A. (2000) In Search of the Ideal Nose. *Plastic and Reconstructive Surgery*, 105, 2561–7.

Chancer, L. S. (1998) *Reconcilable Differences: Confronting Beauty, Pornography, and the Future of Feminism*. Berkeley, University of California Press.

Chapkis, W. (1988) *Beauty Secrets: Women and the Politics of Appearance*. London, Women's Press.

Cheek, J., Shoebridge, J., Willis, E. and Zadoroznyj, M. (1996) Challenging Society: Critical Theory, Foucauldian Postmodernism, and the Poststructuralism of Jacques Derrida. In J. Cheek, J. Shoebridge, E. Willis and M. Zadoroznyj (Eds) *Society and Health: Social Theory for Health Workers*. Melbourne, Longman.

Church, J. (1997) Ownership of the Body. In D. T. Meyers (Ed.) *Feminists Rethinking the Self*. Boulder, Colorado, Westview Press.

Comiskey, C. (2004) Cosmetic Surgery in Paris in 1926: The Case of the Amputated Leg. *Journal of Women's History*, 16, 30–54.

Cook (2007) Factors Influencing Surgeons' Decisions in Elective Cosmetic Surgery Consultations. *Medical Decision Making*, 27, 311–20.

Cooke, K. (1994) *Real Gorgeous: The Truth about Body and Beauty*. St Leonards, Allen and Unwin.

Corrigan, A. and Meredyth, D. (1997) The Body Politic. In K. Pritchard Hughes (Ed.) *Contemporary Australian Feminism 2*. 2nd edn. Melbourne, Longman.

Corrigan, O. (2003) Empty Ethics: The Problem with Informed Consent. *Sociology of Health and Illness*, 25, 768–92.

Cosmetic Surgery Consultants Ltd (2006) Recent Cosmetic Surgery Statistics.

Covino, D. C. (2001) Outside-In: Body, Mind, and Self in the Advertisement of Aesthetic Surgery. *Journal of Popular Culture*, 35, 91–102.

Crawford, Trish (1999) Skin Deep: Cosmetic Surgery Turf War. *Australian Doctor*.

Crerand, C. E., Franklin, M. E. and Sarwer, D. B. (2006) Body Dysmorphic Disorder and Cosmetic Surgery. *Plastic and Reconstructive Surgery*, 118, 167e–80e.

Daniel, A. E., Burn, R. J. and Horarik, S. (1999) Patients' Complaints about Medical Practice. *Medical Journal of Australia*, 170, 598–602.

Darling-Wolf, F. (1997) Framing the Breast Implant Controversy: A Feminist Critique. *Journal of Communication Inquiry*, 21, 77–97.

Davis, D. and Vernon, M. L. (2002) Sculpting the Body Beautiful: Attachment Style, Neuroticism, and Use of Cosmetic Surgeries. *Sex Roles*, 47, 129–38.

Davis, K. (1988) *Power Under the Microscope*. Dordrecht, the Netherlands, Foris Publications.

Davis, K. (1993) Cultural Dopes and She-Devils: Cosmetic Surgery as Ideological Dilemma. In S. Fisher and K. Davis (Eds) *Negotiating at the Margins: The Gendered Discourses of Power and Resistance*. New Brunswick, Rutgers University Press.

Davis, K. (1995) *Reshaping the Female Body: The Dilemma of Cosmetic Surgery*. NY, Routledge.

Davis, K. (1996) From Objectified Body to Embodied Subject: A Biographical Approach to Cosmetic Surgery. In S. Wilkinson (Ed.) *Feminist Social Psychologies: International Perspectives*. Buckingham, Open University Press.

Davis, K. (Ed.) (1997) *Embodied Practices: Feminist Perspectives on the Body*. Thousand Oaks, California, Sage Publications.

Davis, K. (1997) Embody-ing Theory: Beyond Modernist and Postmodernist Readings of the Body. In K. Davis (Ed.) *Embodied Practices: Feminist Perspectives on the Body*. Thousand Oaks, California, Sage Publications.

Davis, K. (1997) 'My Body is My Art': Cosmetic Surgery as Feminist Utopia? In K. Davis (Ed.) *Embodied Practices: Feminist Perspectives on the Body*. Thousand Oaks, California, Sage Publications.

Davis, K. (1998) The Rhetoric of Cosmetic Surgery: Luxury or Welfare? In E. Parens (Ed.) *Enahancing Human Traits*. Washington DC, Georgetown University Press.

Davis, K. (1999) Cosmetic Surgery in a Different Voice: The Case of Madam Noel. *Women's Studies International Forum*, 22, 473–88.

Davis, K. (2002) A Dubious Equality: Men, Women and Cosmetic Surgery. *Body and Society*, 8, 49–65.

Davis, K. (2003) *Dubious Equalities and Embodied Differences: Cultural Studies on Cosmetic Surgery*. Lanham, Rowman and Littlefield Publishers.

Davis, K. (2003) Surgical Passing: Or Why Michael Jackson's Nose Makes 'Us' Uneasy. *Feminist Theory*, 4, 73–92.

Davis, K. and Fisher, S. (1993) Power and the Female Subject. In S. Fisher and K. Davis (Ed.) *Negotiating at the Margins: The Gender Discourses of Power and Resistance*. New Brunswick, Rutgers University Press.

Davis, N. Z. and Farge, A. (Eds) (1993) *A History of Women in the West: III. Renaissance and Enlightenment Paradoxes*. Cambridge, MA, Harvard University Press.

Denzin, N. K. and Lincoln, Y. S. (1994) *Handbook of Qualitative Research*. Thousand Oaks, Sage.

Dinnerstein, M. and Weitz, R. (1998) Jane Fonda, Barbara Bush and Other Aging Bodies. In R. Weitz (Ed.) *The Politics of Women's Bodies: Sexuality, Appearance and Behavior*. NY, Open University Press.

Diprose, R. (1994) *The Bodies of Women: Ethics, Embodiment and Sexual Difference*. London, Routledge.

Doniger, W. (2000) The Mythology of the Face-lift. *Social Research*, 67, 99–125.

Doyal, L. (1995) *What Makes Women Sick: Gender and the Political Economy of Health*. Houndmills, Palgrave Macmillan.

Druss, R. (1973) Changes in Body Image Following Augmentation Breast Surgery. *International Journal of Psychoanalytic Psychotherapy*, 2, 248–56.

Dugas, B. B. (1999) The Good Old Days: A Look Back at Cosmetic Surgery. *Plastic Surgical Nursing*, 19, 74–6, 106.

Dull, D. and West, C. (1991) Accounting for Cosmetic Surgery: The Accomplishment of Gender. *Social Problems*, 38, 801–17.

Dunofsky, M. (1997) Psychological Characteristics of Women who Undergo Single and Multiple Cosmetic Surgeries. *Annals of Plastic Surgery*, 39, 223–8.

Durkin, S. J. and Paxton, S. J. (2002) Predictors of Vulnerability to Reduced Body Image Satisfaction and Psychological Wellbeing in Response to Exposure to Idealized Female Media Images in Adolescent Girls. *Journal of Psychosomatic Research*, 53, 995–1005.

Edelmann, R. J. (2000) *Psychosocial Aspects of the Health Care Process*. Harlow, England, Prentice Hall.

Elix, J. and Lambert, J. (1999) Survey of Consumers of Cosmetic Surgery: A Report Undertaken for the Cosmetic Surgery Inquiry. Sydney, Community Solutions, 179 Sydney Rd. Fairlight NSW 2094.

Elliott, C. (2003) *Better than Well: American Medicine meets the American Dream.* NY, W.W. Norton.

Emanuel, E. J. and Emanuel, L. L. (1999) Four Models of the Physician–Patient Relationship. In J. D. Arras and B. Steinbock (Eds) *Ethical Issues in Modern Medicine.* 5th edn. Mountain View, CA, Mayfield Publishing Company.

English, B. G., Solomon, M. R. and Ashmore, R. D. (1994) Beauty *Before* the Eyes of Beholders: The Cultural Encoding of Beauty Types in Magazines Advertising and Music Television. *Journal of Advertising*, XXXIII, 49–64.

Ensel, A. (1997) Cutting in Measures: Cosmetic Surgery and the Relationship Between the Sexes in Wesern Medicine. *Curare*, 11, 231–4.

Etcoff, N. (1999) *Survival of the Prettiest: The Science of Beauty.* NY, Doubleday.

Fairhurst, E. (1998) 'Growing Old Gracefully' as Opposed to 'Mutton Dressed as Lamb': The Social Construction of Recognising Older Women. In S. Nettleton and J. Watson (Ed.) *The Body in Everyday Life.* London, Routledge.

Fallon, A. (1990) Culture in the Mirror: Sociocultural Determinants of Body Image. In T. F. Cash and T. Pruzinsky (Eds) *Body Images: Development, Deviance and Change.* NY, The Guildford Press.

Fallon, A. (1994) Body Image and the Regulation of Body Weight. In V. J. Adesso, D. M. Fleming and R. Fleming (Eds) *Psychological Perspectives on Women's Health.* Washington DC, Taylor and Francis.

Fallon, P., Katzman, M. A. and Wooley, S. C. (Eds) (1994) *Feminist Perspectives on Eating Disorders.* NY, The Guildford Press.

Featherstone, M. (1999) Body Modification: An Introduction. *Body and Society*, 5, 1–13.

Featherstone, M. (Ed.) (2000) *Body Modification.* London, Sage Publications.

Feingold, A. and Mazzella, R. (1998) Gender Differences in Body Image are Increasing. *Psychological Science*, 9, 190–5.

Fielding, H. (1996) Grounding Agency in Depth: The Implications of Merleau-Ponty's Thought for the Politics of Feminism. *Human Studies*, 19, 175–84.

Fisher, S. (1990) The Evolution of Psychological Concepts about the Body. In T. F. Cash and T. Pruzinsky (Eds) *Body Images: Development, Deviance and Change.* NY, The Guildford Press.

Fisher, S. and Davis, K. (Eds) (1993) *Negotiating at the Margins: The Gendered Discourses of Power and Resistance.* New Brunswick, Rutgers University Press.

Foucault, M. (1972) *The Archaeology of Knowledge and the Discourse on Language.* NY, Pantheon Books.

Foucault, M. (1973) *The Birth of the Clinic: An Archeology of Medical Perception.* NY, Vintage Books, Random House.

Foucault, M. (1979) *Discipline and Punish: The Birth of the Prison.* NY, Vintage.

Frank, A. W. (1998) How Images Shape Bodies. *Body and Society*, 4, 101–12.

Frank, K. (2006) Agency.

Fraser, S. (2001) Woman-Made Women: Mobilisation of Nature in Feminist Accounts of Cosmetic Surgery. *Hecate*, 27, 115–32.

Fraser, S. (2003a) The Agent Within: Agency Repertoires in Medical Discourse on Cosmetic Surgery. *Australian Feminist Studies*, 18, 27–44.

Fraser, S. (2003b) *Cosmetic Surgery, Gender and Culture.* Houndmills, Basingstoke; NY, Palgrave Macmillan.

Friedson, E. (1973) *Profession of Medicine: A Study of the Sociology of Applied Knowledge.* NY, Dodd, Mead and Co.

Frost, L. (1999) 'Doing Looks': Women, Appearance and Mental Health. In J. Arthurs and J. Grimshaw (Ed.) *Women's Bodies: Disciplines and Transgression.* London, Cassell.

Gabe, J. (1995) Health, Medicine and Risk: The Need for a Sociological Approach. In J. Gabe (Ed.) *Medicine, Health and Risk.* Oxford, Blackwell Publishers.

Gadsden, G. Y. (2000) The Male Voice in Women's Magazines. *Gender Issues,* Spring, 49–57.

Gafni, A., Charles, C. and Whelan, T. (1998) The Physician-Patient Encounter: The Physician as a Perfect Agent for the Patient *versus* the Informed Treatment Decision-Making Model. *Social Science and Medicine,* 47, 347–54.

Gagne, P. and Mcgaughey, D. (2002) Designing Women: Cultural Hegemony and the Exercise of Power among Women who have undergone Elective Mammoplasty. *Gender and Society,* 16, 814–38.

Gale Group. *Cosmetic Surgery Times.* Duluth, MN, Advanstar Communications.

Ganellen, R. J. (1996) *Integrating the Rorschach and the MMPI-2 in Personality Assessment.* Mahwah, NJ, Lawrence Erlbaum Associates.

Gear, R. (2001) All those Nasty Womanly Things: Women Artists, Technology and the Monstrous-Feminine. *Women's Studies International Forum,* 24, 321–33.

Gergen, M. and Gergen, K. (1993) Narratives of the Gendered Body in Popular Autobiography. In R. Josselson and A. Lieblich (Ed.) *The Narrative Study of Lives.* Newbury Park, CA, Sage.

Giddens, A. (1991) *Modernity and Self Identity: Self and Society in the Late Modern Age.* Oxford, Polity Press.

Gillespie, M. A. (1998) Mirror Mirror. In R. Weitz (Ed.) *The Politics of Women's Bodies:Sexuality, Appearance and Behaviour.* NY, Open University Press.

Gillespie, R. (1996) Women, the Body and Brand Extension in Medicine: Cosmetic Surgery and the Paradox of Choice. *Women and Health,* 24, 69–85.

Gilman, S. L. (1998) *Creating Beauty to Cure the Soul: Race and Psychology in the Shaping of Aesthetic Surgery.* Durham, NC, Duke University Press.

Gilman, S. L. (1999) *Making the Body Beautiful: A Cultural History of Cosmetic Surgery.* Princeton, NJ, Princeton University Press.

Gimlin, D. (2000) Cosmetic Surgery: Beauty as Commodity. *Qualitative Sociology,* 23, 77–98.

Gjerberg, E. (2001) Medical Women – Towards Full Integration? An Analysis of the Specialty Choices made by Two Cohorts of Norwegian Doctors. *Social Science and Medicine,* 52, 331–43.

Gjerberg, E. (2003) Women Doctors in Norway: The Challenging Balance between Career and Family Life. *Social Science and Medicine,* 57, 1327–41.

Goin, M. K., Burgoyne, M. D., Goin, J. M. and Staples, F. R. (1980) A Prospective Psychological Study of 50 Female Face-lift Patients. *Plastic and Reconstructive Surgery,* 436–42.

Goin, M. K. and Goin, J. M. (1986) Psychological Effects of Aesthetic Facial Surgery. *Advances Psychosomatic Medicine,* 15, 84–108.

Goldman, H. G. (1993) *Fanny Brice: The Original Funny Girl.* NY, Oxford University Press.

Goodman, M. (1994) Social, Psychological and Developmental Factors in Women's Receptivity to Cosmetic Surgery. *Journal of Aging Studies,* 8, 375–96.

Goodman, M. (1996) Culture, Cohort, and Cosmetic Surgery. *Journal of Women and Aging,* 8, 55–73.

Gorney, M. and Martello, J. (1999) Patient Selection Criteria. *Clinics of Plastic Surgery*, 26, 37–40.

Grazer, F. M. and de Jong, R. H. (2000) Fatal Outcomes from Liposuction: Census Survey of Cosmetic Surgeons. *Plastic and Reconstructive Surgery*, 105, 436–46.

Green, E., Thompson, D. and Griffiths, F. (2002) Narratives of Risk: Women at Midlife, Medical 'experts' and Health Technologies. *Health, Risk and Society*, 4, 273–86.

Greico, S. F. M. (1993) The Body, Appearance and Sexuality. In N. Z. Davis and A. Farge (Eds) *A History of Women in the West: III Renaissance and Enlightenment Paradoxes.* Cambridge, MA, Harvard University Press.

Groesz, L. M., Levine, M. P. and Murnen, S. K. T. (2002) The Effects of Experimental Presentation of Thin Images on Body Satisfaction: A Meta-Analytic Review. *International Journal of Eating Disorders*, 31, 1–16.

Grogan, S. (1999) *Body Image: Understanding Body Dissatisfaction in Men, Women and Children.* London, Routledge.

Grogan, S. and Wainwright, N. (1996) Growing Up in the Culture of Slenderness: Girls' Experience of Body Dissatisfaction. *Women's Studies International Forum*, 19, 665–73.

Grogan, S., Williams, Z. and Conner, M. (1996) The Effect of Viewing Same-Gender Photographic Models on Body-Esteem. *Psychology of Women Quarterly*, 20, 569–75.

Grossbart, T. A., Sarwer, D. B., Pertschuk, M. J., Wadden, T. A. and Whitaker, L. A. (1999) Cosmetic Surgery: Surgical Tools – Psychological Goals. *Seminars in Cutaneous Medicine and Surgery*, 18, 101–11.

Grosz, E. (1987) Notes Towards a Corporeal Feminism. *Australian Feminist Studies*, 5, 1–16.

Grosz, E. (Ed.) (1991) *Feminism and the body.* Bloomington, IN, Indiana University Press.

Grosz, E. (1994) *Volatile Bodies: Towards a Corporeal Feminism.* Bloomington, IN, Indiana University Press.

Haiken, E. (1997) *Venus Envy: A History of Cosmetic Surgery.* Baltimore, John Hopkins University Press.

Haiken, E. (2000) The Making of the Modern Face: Cosmetic Surgery. *Social Research*, 67, 81–97.

Halprin, S. (1995) *'Look at my Ugly Face!': Myths and Musings on Beauty and Other Perilous Obsessions with Women's Appearance.* NY, Viking.

Hancock, P., Hughes, B., Tyler, M., Jagger, E., Paterson, K., Russell, R., Tulle-Winton, E. and Tyler M. (2000) *The Body, Culture and Society: An Introduction.* Milton Keynes, Open University Press.

Hansen, J. and Reed, E. (Eds) (1986) *Cosmetics, Fashions and the Exploitation of Women.* NY, Pathfinder Press.

Haraway, D. J. (1991) *Simians, Cyborgs, and Women: The Reinvention of Nature.* NY, Routledge.

Harris, D. (1989) The Benefits and Hazards of Cosmetic Surgery. *British Journal of Hospital Medicine*, 41, 540–5.

Hasan, J. S. (2000) Psychological Issues in Cosmetic Surgery: A Functional Overview. *Annals of Plastic Surgery*, 44, 89–96.

Hayflick, L. (2000) The Future of Ageing. *Nature*, 408, 267–9.

Hayry, H. (1991) *The Limits of Medical Paternalism.* London, Routledge.

Health Complaints Commission, N. S. W. (1999) The Cosmetic Surgery Report: Report to the NSW Minister for Health. Sydney, Health Complaints Commission.

Health Services Commission (2002) Health Service Commissioner Annual Report. Melbourne.

Heinberg, L. J. and Thompson, J. K. (1995) Body Images and Televised Images of Thinness: A Controlled Laboratory Investigation. *Journal of Social and Clinical Psychology*, 14, 325–38.

Henning, M. (1999) Don't Touch Me (I'm Electric): On Gender and Sensation in Modernity. In J. Arthurs and J. Grimshaw (Eds.) *Women's Bodies: Discipline and Transgression*. London, Cassell.

Hesse-Biber, S. (1996) *Am I thin Enough Yet? The Cult of Thinness and the Commercialization of Identity*. NY, Routledge.

Hesse-Biber, S. (2006) The Mass Marketing of Disordered Eating and Eating Disorders: The Social Psychology of Women, Thinness and Culture. *Women's Studies International Forum*, 29, 208–24.

Heyes, C. J. (2007) Cosmetic Surgery and the Televisual Makeover. *Feminist Media Studies*, 7, 17–32.

Heyes, C. J. (2007) Normalisation and the Psychic Life of Cosmetic Surgery. *Australian Feminist Studies*, 22, 55–71.

Heyes, C. J. (2007) *Self-Transformations: Foucault, Ethics, and Normalized Bodies*. NY, Oxford University Press.

Hillier, J. (1997) Foucault's Gaze. In O'Farrell, C. (Ed.) *Foucault: The Legacy*. Brisbane, Queensland University of Technology Publishing.

Hoeyberghs, J. L. (1999) Cosmetic Surgery. *British Medical Journal*, 318, 512–16.

Holliday, R. and Taylor, J. S. (2006) Aesthetic Surgery as False Beauty. *Feminist Theory*, 7(2), 179–95.

Hovi, S-L., Hemminki, E. and Swan, S. H. (1999) Cosmetic and Postmasectomy Breast Implants: Finnish Women's Experiences. *Journal of Women's Health and Gender Based Medicine*, 8, 933–9.

Howson, A. (1998) Embodied Obligation: The Female Body and Health Surveillance. In S. Nettleton and J. Watson (Ed.) *The Body in Everyday Life*. London, Routledge.

Huon, G., Morris, S. and Brown, L. (1990) Differences between Male and Female Preferences for Female Body Size. *Australian Psychologist*, 25, 314–17.

Hurd Clarke, L. and Griffin, M. (2007) The Body Natural and the Body Unnatural: Beauty Work and Aging. *Journal of Aging Studies*, 21, 187–201.

Hussain, R., Schofield, M. and Loxton, D. (2002) Cosmetic Surgery History and Health Service Use in Midlife: Women's Health Australia. *Medical Journal of Australia*, 176, 576–9.

Hyde, P. (2000) Managing Bodies – Managing Relationships: The Popular Media and the Social Construction of Women's Bodies and Social Roles from the 1930's to the 1950's. *Journal of Sociology*, 36, 157–71.

Illich, I. (1976) *Medical Nemesis: The Expropriation of Health*. NY, Pantheon.

Ishigooka, J., Mitsuhiro, I., Makihiko, S., Yoshitsuna, F., Mitsukuni, M. and Sadanori, M. (1998) Demographic Features of Patients Seeking Cosmetic Surgery. *Psychiatry and Clinical Neurosciences*, 52, 283–7.

Jacobson, N. (2000) *Cleavage: Technology, Controversy and the Ironies of the Man-made Breast*. New Brunswick, Rutgers University Press.

Jaggar, A. M. (Ed.) (1994) *Living with Contradictions*. Boulder, CO, Westview Press.

Jeffereys, S. (2000) 'Body Art' and Social Status: Cutting, Tattooing and Piercing from a Feminist Perspective. *Feminism and Psychology*, 10, 409–29.

Johansson, E. E., Hamberg, K., Westman, G. and Lindgren, G. (1999) The Meanings of Pain: An Exploration of Women's Descriptions of Symptoms. *Social Science and Medicine*, 48, 1791–802.

Jones, M. (2008) Makeover Culture's Dark Side: Breasts, Death and Lolo Ferrari. *Body and Society*, 14, 89–104.

Jones, M. (2004) Architecture of the Body: Cosmetic Surgery and Postmodern Space. *Space and Culture*, 7, 90–101.

Jones, M. (2008) *Skintight: An Anatomy of Cosmetic Surgery*. Oxford, Berg.

Jordan, J. W. (2004) The Rhetorical Limits of the 'Plastic Body'. *Quarterly Journal of Speech*, 90, 327–58.

Josselson, R. and Lieblich, A. (Eds) (1993) *The Narrative Study of Lives*. Newbury Park, CA, Sage.

Kalick, S. M. (1979) Aesthetic Surgery: How it Affects the Way Patients are Perceived by Others. *Annnals of Plastic Surgery*, 2, 128–34.

Kaplan, E. A. (1996) *Looking for the Other: Feminism and the Imperial Gaze*. NY, Routledge.

Kapp, M. B. (2007) Patient Autonomy in the Age of Consumer-Driven Health Care: Informed Consent and Informed Choice. *Journal of Legal Medicine*, 28, 91–117.

Katz, S., Kravetz, S. and Marks, Y. (1997) Parent's and Doctor's Attitudes toward Plastic Facial Surgery for Persons with Down Syndrome. *Journal of Intellectual and Developmental Disability*, 22, 265–73.

Kaw, E. (1994) 'Opening' Faces. In N. Sault (Ed.) *Many Mirrors: Body Image and Social Relations*. New Brunswick, Rutgers University Press.

Kaw, E. (1998) Medicalization of Racial Features: Asian-American Women and Cosmetic Surgery. In R. Weitz (Ed.) *The Politics of Women's Bodies: Sexuality, Appearance and Beahavior*. NY, Open University Press.

Kenardy, J., Brown, W. J. and Vogt, E. (2001) Dieting and Health in Young Australian Women. *European Eating Disorders Review*, 9, 242–54.

Kilbourne, J. (1994) Still Killing us Softly: Advertising and the Obsession with Thinness. In P. Fallon, M. A. Katzman and S.C. Wooley (Eds) *Feminist Perspectives on Eating Disorders*. NY, The Guildford Press.

Kirkland, A. and Tong, R. (1996) Working within Contradictions: The Possibility of Feminist Cosmetic Surgery. *The Journal of Clinical Ethics*, 7, 151–9.

Klassen, A., Fitzpatrick, R., Jenkinson, C. and Goodacre, T. (1996) Should Breast Reduction Surgery be Rationed? A Comparison of the Health Status of Patients before and after Treatment: Postal Questionnaire Survey. *British Medical Journal*, 313, 454–7.

Koch, R. J., Chavez, A., Dagum, P. and Newman, J. P. (1998) Advantages and Disadvantages of Computer Imaging in Cosmetic Surgery. *Dermatological Surgery*, 24, 195–8.

Komesaroff, P. (1995) Introduction: Postmodern ethics? In P. Komesaroff (Ed.) *Troubled Bodies: Critical Perspectives on Postmodernism, Medical Ethics, and the Body*. Melbourne, Melbourne University Press.

Komesaroff, P. (Ed.) (1995) *Troubled Bodies: Critical Perspectives on Postmodernism, Medical Ethics, and the Body*. Melbourne, Melbourne University Press.

Koval, R. (1986) *Eating Your Heart Out: Food, Shape and the Body Industry*. Sydney, Penguin Books.

Krieger, L. M. (2002) Discount Cosmetic Surgery: Industry Trends and the Strategies for Success. *Plastic and Reconstructive Surgery*, 110, 614–19.

Lakkis, J. and Ricciardelli, L. A. (1999) Role of Sexual Orientation and Gender-Related Traits in Disordered Eating. *Sex Roles*, 41, 1–16.

Lamb, C. S., Jackson, L. A., Cassiday, P. B. and Priest, D. J. (1993) Body Figure Preferences of Men and Women: A Comparison of Two Generations. *Sex Roles*, 28, 345–58.

Lane, K. (1995) The Medical Model of the Body as a Site of Risk: A Case Study of Childbirth. In J. Gabe (Ed.) *Medicine, Health and Risk: Sociological Approaches*. Oxford, Blackwell.

Lee, C. (1998) *Women's Health: Psychological and Social Perspectives*. London, Sage Publications.

Leinert, T. (1998) Women's Self-Starvation, Cosmetic Surgery and Transsexualism. *Feminism and Psychology*, 8, 245–50.

Levinson, W., Roter, D. L., Mullooly, J. P., Dull, V. T. and Frankel, R. M. (1997) Physician–Patient Communication. The Relationship with Malpractice Claims among Primary Care Physicians and Surgeons. *Journal of the American Medical Association*, 277, 553–9.

Lindenmeyer, A. (1999) Postmodern Concepts of the Body in Jeanette Winterson's 'Written on the Body'. *Feminist Review*, 61, 48–63.

Lingis, A. (1996) The Body Postured and Dissolute. In V. M. Fóti (Ed.) *Merleau-Ponty: Difference, Materiality, Painting*. NJ, Humanities Press.

Little, M. O. (1998) Cosmetic Surgery, Suspect Norms and the Ethics of Complicity. In E. Parens (Ed.) *Enhancing Human Traits*. Washington DC, Georgetown University Press.

Lloyd, M. and Thacker, A. (1997) Foucault's Ethics and Politics: A Strategy for Feminism. In M. Lloyd and A. Thacker (Eds) *The Impact of Michael Foucault on the Social Sciences and Humanities*. Basingstoke, Palgrave Macmillan.

Lloyd, M. and Thacker, A. (Eds) (1997) *The Impact of Michael Foucault on the Social Sciences and Humanities*. Basingstoke, Palgrave Macmillan.

Lock, M. (1998) Anamolous Ageing: Managing the Postmenopausal Body. *Body and Society*, 4, 35–61.

Lupton, D. (1994) *Medicine as Culture: Illness, Disease, and the Body in Western Societies*. London, Sage.

Lupton, D. (1995) *The Imperative of Health: Public Health and the Regulated Body*. London, Thousand Oaks; CA, Sage Publications.

Lupton, D. (1997) Consumerism, Reflexivity and the Medical Encounter. *Social Science and Medicine*, 45, 373–81.

Magill, K. (1997) Surveillane-Free Subjects. In M. Lloyd and A. Thacker (Eds) *The Impact of Michael Foucault on the Social Sciences and Humanities*. Basingstoke, Palgrave Macmillan.

Markus, R. F. (1999) Lasers: More than a Cosmetic Tool. *Journal of Dermatological Treatment*, 11, 117–24.

Marshall, D. (1998) *Your Face in their Hands: Cosmetic Surgery – Risks and Rewards*. Melbourne, Hill of Content.

Marshall, H. (1996) Our Bodies Our Selves: Why We should Add Old Fashioned Empirical Phenomenology to the New Theories of the Body. *Women's Studies International Forum*, 19, 253–65.

Marshall, P. A. (1996) Boundary Crossings: Gender and Power in Clinical Ethics Consultations. In C. F. Sargent and C. B. Brettell (Eds) *Gender and Health: An International Perspective*. New Jersey, Prentice Hall.

Marwick, A. (1988) *Beauty in History: Society, Politics and Personal Appearance c.1500 to the Present*. London: Thames and Hudson.

Mason, J. (2002) *Qualitative Researching*. London, Sage.

Matthews Grieco, S. F. (1993) The Body, Appearance, and Sexuality. In N. Z. Davis and A. Farge (Eds) *A History of Women in the West: Renaissance and Enlightenment Paradoxes*. Cambridge, MA, The Belknap Press of Harvard University Press.

Mazzeo, J. J. (2007) Effects of a Reality TV Cosmetic Surgery Makeover Program on Eating Disordered Attitudes and Behaviors. *Eating Behaviors*, 8, 390–7.

McCabe, M. and Ricciardelli, L. (1999) Socio-Cultural Influences on Body Image and Body Change Strategies among Adolescent Boys. Presented at The Body Culture Conference: Challenging Current Approaches to Physical Activity, Healthy Eating, Sexual Health and Body Image. Melbourne, July.

McClean, S. (1989) *A Patient's Right to Know: Information Disclosure, the Doctor and the Law*. Aldershot, Dartmouth.

Mcgrath, M. H. and Mukerji, S. (2000) Plastic Surgery and the Teenage Patient. *Journal of Pediatric and Adolescent Gynaecology*, 13, 105–18.

Mclaughlin, J. K., Wise, T. N. and Lipworth, L. (2004) Increased Risk of Suicide among Patients with Breast Implants: Do the Epidemiologic Data Support Psychiatric Consultation? *Psychosomatics*, 45, 277–80.

Mclaughlin, T. L. and Goulet, N. (1999) Gender Advertisements Aimed at African Americans: A Comparison to their Occurrence in Magazines Aimed at Caucasians. *Sex Roles*, 40, 61–71.

Mcmillan, E. (1987) Female Difference in the Texts of Merleau-Ponty. *Philosophy Today*, 31, 359–66.

Mcnay, L. (1991) The Foucauldian Body and the Exclusion of Experience. *Feminism and the Body, Hypatia Special Issue*, 6.

Mcnay, L. (2000) *Gender and Agency: Reconfiguring the Subject in Feminist Social Theory*. Cambridge, Polity Press.

McNulty, B. (1998) The Bust. *Good Medicine*.

Mead, N., Bower, P. and Hann, M. (2002) The Impact of General Practitioners' Patient-Centredness on Patients' Post-Consultation Satisfaction and Enablement. *Social Science and Medicine*, 55, 283–99.

Medical Practitioners Board of Victoria Advertising Guidelines for Registered Medical Practitioners. Melbourne, Medical Practitioners Board of Victoria.

Medical Practice Act, Australia.(1994)

Meisler, J. G. (2000) Toward Optimal Health: The Experts Discuss Cosmetic Surgery. *Journal of Women's Health and Gender Based Medicine*, 9, 13–18.

Mellor, P. A. and Shilling, C. (1997) *Re-forming the Body: Religion, Community and Modernity*. London, Sage Publications.

Mendelsohn, R. S. (1982) *Male Practice: How Doctors Manipulate Women*. Chicago, Contemporary Books.

Meningaud, J. P., Srvant, J. M., Herve, C. and Bertrand, J. C. (2000) Ethics and Aims of Cosmetic Surgery: A Contribution from an Analysis of Claims after Minor Damage. *Medicine and Law*, 19, 237–52.

Merleau-Ponty, M. (1992) Being and Having (1936). In M. B. Smith translated *Texts and Dialogues*. New Jersey, Humanities Press International Inc.

Meyers, D. T. (Ed.) (1997) *Feminists Rethinking the Self*. Boulder, CO, Westview Press.

Miles, A. (1998) Women's Bodies, Women's Selves: Illness Narratives and the 'Andean' Body. *Body and Society*, 4, 1–19.

Miles, M. B. and Huberman, A. M. (1994) *Qualitative Data Analysis*. Thousand Oaks, Sage Publications.

Miller, F. G., Brody, H. and Chung, K. C. (2000) Cosmetic Surgery and the Internal Morality of Medicine. *Cambridge Quarterly of Health Ethics*, 9, 353–64.

Miranda, D. (2007) They're So Beautiful they could just Die. *The Sydney Morning Herald*. 2nd February, Sydney.

Monder, J. J. M. (2007) After Our Bodies Ourselves. *Psychoanalytic psychology*, 24, 384–94.

Morgan, D. H. J. and Scott, S. (1993) Bodies in a Social Landscape. In S. Scott and D. H. J. Morgan (Eds) *Body Matters: Essays on the Sociology of the Body*. London, The Falmer Press.

Morgan, K. P. (1991) Women and the Knife: Cosmetic Surgery and the Colonization of Women's Bodies. *Hypatia*, 5, 25–53.

Mudge, T. J. and Dashwood, D. A. (2002) A Change in the Make-Up of Medicine. *Medical Journal of Australia*, 176, 569–70.

Murad, A. and Dover, J. (2001) On Beauty: Evolution, Psychosocial Considerations and Surgical Enhancement. *Archives of Dermatology*, 137, 795–807.

Myers, P. N. and Biocca, F. (1992) The Elastic Body Image: The Effect of Television Advertising and Programming on Body Image Distortions in Young Women. *Journal of Communication*, 42, 108–33.

Negrin, L. (1999) The Self as Image: A Critical Appraisal of Postmodern Theories of Fashion. *Theory, Culture and Society*, 16, 99–118.

Negrin, L. (2002) Cosmetic Surgery and the Eclipse of Identity. *Body and Society*, 8, 21–42.

Nettleton, S. (1995) *The Sociology of Health and Illness*. Cambridge, Polity Press.

Nettleton, S. (1996) Women and the New Paradigm of Health and Medicine. *Critical Social Policy*, 48, 33–53.

Nettleton, S. and Watson, J. (Eds) (1998) *The Body in Everyday Life*. London, Routledge.

Nettleton, S. and Watson, J. (1998) The Body in Everyday Life: An Introduction. In S. Nettleton and J. Watson (Eds) *The Body in Everday Life*. London, Routledge.

New South Wales. Health Care Complaints Commission and Cornwall A (1999) The Cosmetic Surgery Report: Report to the NSW Minister for Health. Strawberry Hills, N.S.W., The Commission.

Nisselle, P. (1999) Managing Risk in Medical Practice. *Journal of Law and Medicine*, 7, 130–9.

O'Farrell, C. (Ed.) (1997) *Foucault: The Legacy*. Brisbane, Queensland University of Technology Publishing.

Oakley, A. (1980) *Women Confined: Towards a Sociology of Childbirth*. Oxford, Martin Robertson.

Oakley, A. (1998) Science, Gender and Women's Liberation: An Argument Against Postmodernism. *Women's Studies International Forum*, 21, 133–46.

Oakley, A. (2000) *Experiments in Knowing: Gender and Method in the Social Sciences*. Cambridge, Polity Press.

Oguzer, F., Tunkali, D. and Guler-Gursu, K. (1998) Life Satisfaction, Self Esteem and Body Image: A Psychological Evaluation of Aesthetic and Reconstructive Surgery Candidates. *Aesthetic Plastic Surgery*, 22, 412–19.

Oinas, E. (1998) Medicalisation by Whom? Accounts of Menstruation conveyed by Young Women and Medical Experts in Medical Advisory Columns. *Sociology of Health and Illness*, 20, 52–70.

Olds, T. and Norton, K. (1999) How to Become a Supermodel. Presented at The Body Culture Conference: Challenging Current Approaches to Physical Activity, Healthy Eating, Sexual Health and Body Image. Melbourne, July.

Ong, L. M. L., de Haes, J. C. J. M., Hoos, A. M. and Lammes, F. B. (1995) Doctor–Patient Communication: A Review of the Literature. *Social Science and Medicine*, 40, 903–18.

Orbach, S. (1993) *Hunger Strike: The Anorectic's Struggle as a Metaphor for our Age*. London, Penguin.

Orbach, S. (1999) Whose body? The Politics of the Body and the Body Politic. *The Body Culture Conference: Challenging Current Approaches to Physical Activity, Healthy Eating, Sexual Health and Body Image*. Melbourne.

Orton, C. (2002) Regulating Cosmetic Surgery, Editorial. *British Medical Journal*, 324, 1229–30.

Ozgur, F., Tubcali, D. and Guler Gursu, K. (1998) Life Satisfaction, Self-Esteem, and Body Image: A Psychosocial Evaluation of Aesthetic and Reconstructive Surgery Candidates. *Aesthetic Plastic Surgery*, 22, 412–19.

Paquette, M. C. and Raine, K. (2004) Sociocultural Context of Women's Body Image. *Social Science and Medicine*, 59, 1047–58.

Parens, E. (Ed.) (1998) *Enhancing Human Traits*. Washington DC, Gorgetown University Press.

Parker, L. S. (1993) Social Justice, Federal Paternalism, and Feminism: Breast Implants in the Cultural Context of Female Beauty. *Kennedy Institute of Ethics Journal*, 3, 57–76.

Parker, R. (2006) Negotiating the 'Normal' Body: Women, Doctors and Cosmetic Surgery. *The Australian Sociological Association Annual Conference*. Perth, December.

Parry, O. and Pill, R. (1994) 'I Don't Tell him How to Live his Life': The Doctor/Patient Encounter as an Educational Context. In M. Bloor and P. Taraborrelli (Eds) *Qualitative Studies in Health and Medicine*. Aldershot, Avebury.

Parsons, T. (Ed.) (1949) *Essays in Sociological Theory*. Glencoe, Free Press.

Parsons, T. (1949) The Professions and Social Structure. In T. Parsons (Ed.) *Essays in Sociological Theory*. Glencoe, Free Press.

Paxton, S. and Durkin, S. J. (1999) Predictors of Vulnerability in Response to Exposure to Media in Adolescent Girls. Presented at The Body Culture Conference: Challenging Current Approaches to Physical Activity, Healthy Eating, Sexual Health and Body Image. Melbourne, July.

Paxton, S. J. and Phythian, K. (1999) Body Image, Self-esteem, and Health Status in Middle and Later Adulthood. *Australian Psychologist*, 34, 116–21.

Pecore, V. P. (2002) The Culture of Surveillance. *Qualitative Sociology*, 25, 345–58.

Pedwell, C. (2008) Weaving Relational Webs: Theorizing Cultural Difference and Embodied Practice. *Feminist Theory*, 9, 87–107.

Peerson, A. (1995) Foucault and Modern Medicine. *Nursing Inquiry*, 2, 106–14.

Peiss, K. (1990) Making Faces: The Cosmetic Industry and the Cultural Construction of Gender 1890–1930. *Genders*, Issue 7, 143–69.

Petschuck, M. J., Sarwer, D. B., Wadden, T. A. and Whitaker, L. A. (1998) Body Image Dissatisfaction in Male Cosmetic Surgery Patients. *Aesthetic Plastic surgery*, 22, 20–24.

Phillips, D. (1996) Medical Professional Dominance and Client Satisfaction. *Social Science and Medicine*, 42, 1419–25.

Pirani, C. (12 September 2005). Doctors Face Malpractice Claims Surge. *The Australian*, p. 5.

Pruzinsky, T. (1993) Psychological Factors in Cosmetic Plastic Surgery: Recent Developments in Patient Care. *Plastic Surgical Nursing*, 13, 64–71.

Pruzinsky, T. and Edgerton, M. T. (1990) Body-Image Change in Cosmetic Plastic Surgery. In T. F. Cash and T. Pruzinsky (Eds) *Body Images: Development, Deviance, and Change*. NY, The Guildford Press.

Rainbow, P. (1984) *The Foucault Reader*. NY, Random House.

Rankin, M., Borah, G. L., Perry, A. W. and Wey, P. D. (1998) Quality-of-Life Outcomes after Cosmetic Surgery. *Plastic and Reconstructive Surgery*, 102, 2136–45.

Richards, T. (1990) Chasms in Communication. *British Medical Journal*, 301, 1407–8.

Riessman, C. K. (1998) Women and Medicalization: A New Perspective. In R. Weitz (Ed.) *The Politics of Women's Bodies: Sexuality, Appearance and Behavior.* NY, Open University Press.

Ring, A. L. (2002) Using 'Anti-Ageing' to Market Cosmetic Surgery: Just Good Business, or Another Wrinkle on the Face of Medical Practice? *Medical Journal of Australia*, 176, 597–9.

Roberts, F. D. (1999) *Talking About Treatment: Recommendations for Breast Cancer Adjuvant Therapy.* NY, Oxford University Press.

Rogers, B. O. (1976) The Development of Aesthetic Plastic Surgery: A History. *Aesthetic Plastic Surgery*, 1, 3–24.

Rogers v Whitaker (1992) 175 CLR 479.

Rohrich, R. J. (2000) The Globalization of Cosmetic Surgery: The Pursuit of Excellence. *Plastic and Reconstructive Surgery*, 106, 685–6.

Rohrich, R. J. (2000) The Importance of Cosmetic Plastic Surgery Education: An Evolution. *Plastic and Reconstructive Surgery*, 105, 741–2.

Rohrich, R. J. (2000) The Millenium Cosmetic Surgeon: Who are We and Where are We Going? *Plastic and Reconstructive Surgery*, 105, 225–6.

Rose, N. (1996) *Inventing Our Selves: Psychology, Power and Personhood.* Cambridge and NY, Cambridge University Press.

Rosen, J. C. (1990) Body-Image Disturbance in Eating Disorders. In T. F. Cash and T. Pruzinsky (Ed.) *Body Images: Development, Deviance and Change.* NY, The Guildford Press.

Rosenberg v Percival (2001) 75 ALJR 734.

Rothfield, P. (1995) Bodies and Subjects: Medical Ethics and Feminism. In Komesaroff, P. (Ed.) *Troubled Bodies: Critical Perspectives on Postmodernism, Medical Ethics and the Body.* Melbourne, Melbourne University Press.

Rowbotham, S. (1973) *Hidden from History.* London, Pluto Press.

Rowsell, B., Norris, P., Ryan, K. and Weenink, M. (2000) Assessing and Managing Risk and Uncertainty: Women Living with Breast Implants. *Health, Risk and Society*, 2, 205–18.

Russell, D. (1995) Female Bodies and Food: A Case of Ethics and Psychiatry. In P. Komesaroff (Ed.) *Troubled Bodies: Critical Perspectives on Postmodernism, Medical Ethics, and the Body*. Melbourne, Melbourne University Press.

Russell, D. (1995) *Women, Madness and Medicine*. Cambridge, Polity Press.

Ruzek, S. B., Olesen, V. L. and Clarke, A. E. (Eds) (1997) *Women's Health: Complexities and Differences*. Columbus, Ohio State University Press.

Ryan, V. (2002) The Surgical Fix: Physical Capital, Self-Improvement and the Body Beautiful. *Artlink*, 22, 19–25.

Saks, M. (1995) *Professions and the Public Interest: Medical Power, Altruism and Alternative Medicine*, London, Routledge.

Sarwer, D. B. (1997) The 'Obsessive' Cosmetic Surgery Patient: A Consideration of Body Image Dissatisfaction and Body Dysmorphic Disorder. *Plastic Surgical Nursing*, 17, 193–7, 209.

Sarwer, D. B. and Crerand, C. (2004) Body Image and Cosmetic Medical Treatments. *Body Image*, 1, 99–111.

Sarwer, D. B., Nordmann, J. E. and Herbert, J. D. (2000) Cosmetic Breast Augmentation Surgery: A Critical Overview. *Journal of Women's Health and Gender-Based Medicine*, 9, 843–56.

Sarwer, D. B., Pertschuk, M. J., Wadden, T. A. and Whitaker, L. A. (1998b) Psychological Investigation in Cosmetic Surgery: A Look Back and a Look Ahead. *Plastic and Reconstructive Surgery*, 101, 1136–42.

Sarwer, D. B., Wadden, T. A., Pertschuk, M. J. and Whitaker, L. A. (1998) Body Image Dissatisfaction and Body Dysmorphic Disorder in 100 Cosmetic Surgery Patients. *Plastic and Reconstructive Surgery*, 101, 1644–9.

Sarwer, D. B., Wadden, T. A., Pertschuk, M. J. and Whitaker, L. A. (1998) The Psychology of Cosmetic Surgery: A Review and Reconceptualization. *Clinical Psychology Review*, 18, 1–22.

Sarwer, D. B., Wadden, T. A. and Whitaker, L. A. (2002) An Investigation of Changes in Body Image following Cosmetic Surgery. *Plastic and Reconstructive Surgery*, 109, 363–9.

Sault, N. (Ed.) (1994) *Many Mirrors: Body Image and Social Relations*. New Brunswick, Rutgers University Press.

Schiebinger, L. (2000) *Feminism and the Body*. NY, Open University Press.

Schofield, M., Hussain, R., Loxton, D. and Miller, Z. (2002) Psychosocial and Health Behavioural Covariates of Cosmetic Surgery: Women's Health Australia Study. *Journal of Health Psychology*, 7, 445–57.

Scott, C. E. (1987) The Power of Medicine, the Power of Ethics. *The Journal of Medicine and Philosophy*, 12, 332–50.

Scott, S. and Morgan, D. (Eds) (1993) *Body Matters: Essays on the Sociology of the Body*. London, The Falmer Press.

Seid, R. P. (1994) Too Close to the Bone: The Historical Context for Women's Obsession with Slenderness. In K. W. Fallon (Ed.) *Feminist Perspectives on Eating Disorders*. NY, The Guildford Press.

Sharma, U. and Black, P. (2001) Look Good, Feel Better: Beauty Therapy as Emotional Labour. *Sociology*, 35, 913–31.

Sharp, L. (2000) The Commodification of the Body and its Parts. *Annual Review of Anthropology*, 29, 287–328.

Sheldon, S. and Wilkinson, S. (1998) Female Genital Mutilation and Cosmetic Surgery: Regulating Non-Therapeutic Body Modification. *Bioethics*, 12, 263–85.

Sherwin, S. (1989) Feminist and Medical Ethics: Two Different Approaches to Contextual Ethics. *Hypatia*, 4.

Shiffman, M. A. (2000) Cultural Aspects of the Physician. *International Journal of Cosmetic Surgery and Aesthetic Dermatology*, 2, 169–70.

Shilling, C. (1994) *The Body and Social Theory*. London, Sage.

Silverman, D. (2001) *Interpreting Qualitative Data: Methods for Analysing Talk, Texts and Interaction*. London, Sage Publications.

Simpson, M., Buckman, R., Steward, M., Maguire, P., Lipkin, M., Novack, D. and Till, J. (1991) Doctor–Patient Communication: The Toronto Consensus Statement. *British Medical Journal*, 30, 1385–7.

Skene, L. and Smallwood, R. (2002) Informed Consent: Lessons from Australia. *British Medical Journal*, 324, 39–41.

Spillman, D. M. and Everington, C. (1989) Somatotypes Revisited: Have the Media Changed our Perception of Female Body Image? *Psychological Reports*, 64, 887–90.

Spitzack, C. (1987) Confession and Signification: The Systematic Inscription of Body Conciousness. *The Journal of Medicine and Philosophy*, 12, 356–69.

Spitzack, C. (1988a) The Confession Mirror: Plastic Images for Surgery. *Canadian Journal of Political and Social Theory*, 12, 38–50.

Spitzack, C. (1988b) Under the Knife: Self Knowledge Through Effacement. *Quarterly Journal of Ideology*, 12, 1–18.

Squire, S. (2002) The Personal and the Political: Writing the Theorist's Body. *Australian Feminist Studies*, 17, 55–64.

Strathern, A. (1996) *Body Thoughts*. MI, University of Michigan Press.

Stratton, J. (1996) *The Desirable Body: Cultural Fetishism and the Erotics of Consumption*. Manchester, Manchester University Press.

Sullivan, D. A. (1993) Cosmetic Surgery: Market Dynamics and Medicalization. *Research in the Sociology of Health Care*, 10, 97–115.

Sullivan, D. A. (2001) *Cosmetic Surgery: The Cutting Edge of Commercial Medicine in America*. New Brunswick, Rutgers University Press.

Tait, S. (2007) Television and the Domestication of Cosmetic Surgery. *Feminist Media Studies*, 7, 119–135.

Tan v Benkovic (2000) 51 NSWLR 292.

Tantleff-Dunn, S. (2001) Breast and Chest Size: Ideals and Stereotypes through the 1990s. *Sex Roles*, 45, 231–42.

Taylor, B. W. (2008) The History of Australian Plastic Surgery, www.plasticsurgery.com.au/history. Accessed on 2nd November.

Thesander, M. (1997) *The Feminine Ideal*. London, Reaktion Books.

Thorpe, S. J., Ahmed, B. and Steer, K. (2004) Reasons for Undergoing Cosmetic Surgery: A Retrospective Study. *Sexualities, Evolution and Gender*, 6, 75–96.

Tiggemann, M. and Pennington, B. (1990) The Development of Gender Differences in Body-Size Dissatisfaction. *Australian Psychologist*, 25, 306–13.

Timmermans, D., Molewijk, B., Stiggelbout, A. and Kievit, J. (2004) Different Formats for Communicating Surgical Risks to Patients and the Effect on Choice of Treatment. *Patient Education and Counseling*, 54, 255–63.

Tomycz, N. D. (2006) A Profession Selling Out: Lamenting the Paradigm Shift in Physician Advertising. *Journal of Medical Ethics*, 32, 26–8.

Treichler, P. A., Cartwright, L. and Penley, C. (Eds) (1998) *The Visible Woman: Imaging Technologies, Gender, and Science*. NY, University Press.

Turner, B. S. (1992) *Regulating Bodies: Essays in Medical Sociology*. London, Routledge.

Turner, B. S. (1995) *Medical Power and Social Knowledge*. London, Sage Publications.

Turner, B. S. (1996) *The Body and Society*. London, Sage Publications.

Urla, J. and Swedlund, A. C. (2000) The Anthropometry of Barbie: Unsettling Ideals of the Feminine Body in Popular Culture. In L. Schiebinger (Ed.) *Feminism and the Body*. NY, Oxford University Press.

Van Den Brink-Muinen, A., Van Dulmen, S., Messerli-Rohrbach, V. and Bensing, J. (2002) Do Gender-Dyads have Different Communication Patterns? A Comparative Study in Western-European General Practices. *Patient Education and Counselling*, 48, 253–64.

Van Dulmen, A. M. (2002) Different Perspectives of Doctor Patient in Communication. *International Congress Series*, 1241, 243–48.

Van Manen, M. (1990) *Researching Lived Experience: Human Science for Action Sensitive Pedagogy*. Albany, NY, State University of New York Press.

Veatch, R. M. (Ed.) (1997) *Medical Ethics*. Sudbury, MA, Jones and Bartlett.

Veatch, R. M. (1999) Abandoning Informed Consent. In H. Kuhse and P. Singer (Eds) *Bioethics: An Anthology*. Oxford, Blackwell.

Veatch, R. M. (2000) *The Basics of Bioethics*. New Jersey, Prentice-Hall.

Wallace, A. F. (1982) *The Progress of Plastic Surgery: An Introductory History*. Oxford, Willem A. Meeuws.

Wallace, A. F. (1997) *The History of the British Association of Plastic Surgeons*. London, BAPRAS.

Wardle, J. (1993) Culture and Body Image: Body Perception and Weight Concern. *Journal of Communication and Applied Social psychology*, 3, 377–91.

Warren, V. L. (1992) Feminist Directions in Medical Ethics. In H. B. Holmes and L. M. Purdy (Eds) *Feminist Perspectives in Medical Ethics*. Bloomington, Indiana University Press.

Waterhouse, R. (1993) The Inverted Gaze. In S. Scott and D. Morgan (Eds) *Body Matters: Essays on the Sociology of the Body*. London, The Falmer Press.

Watts, J. (2004) China's Cosmetic Surgery Craze. *The Lancet*, 363, 958.

Webster, F. (2002) Do Bodies Matter? Sex, Gender and Politics. *Australian Feminist Studies*, 17, 191–205.

Weiss, G. (1999) *Body Images: Embodiment and Intercorporeality*. NY, Routledge.

Weitz, R. (2003) *The Politics of Women's Bodies : Sexuality, Appearance, and Behavior*. NY, Oxford University Press.

Welton, D. (1998) *Body and Flesh: A Philosophical Reader*. Cambridge, MA, Blackwell Publishers.

White, D. (1995) Divide and Multiply: Culture and Politics in the New Medical Order. In P. Komesaroff (Ed.) *Troubled Bodies: Critical Perspectives on Postmodernism, Medical Ethics, and the Body*. Melbourne, Melbourne University Press.

Wijsbeck, H. (2000) The Pursuit of Beauty: The Enforcement of Aesthetics or a Freely Adopted Lifestyle? *Journal of Medical Ethics*, 26, 454–8.

Wilkinson, S. (Ed.) (1996) *Feminist Social Psychologies: International Perspectives*. Buckingham, Open University Press.

Witz, A. (2000) Whose Body Matters? Feminist Sociology and the Corporeal Turn in Sociology and Feminism. *Body and Society*, 6, 1–24.

Wolf, N. (1990) *The Beauty Myth: How Images of Beauty are Used against Women.* London, Vintage.

Wolf, S. M. (Ed.) (1996) *Feminism and Bioethics: Beyond Reproduction.* NY, Oxford University Press.

Wolf, S. M. (1996) Introduction: Gender and Feminism in Bioethics. In Wolf, S. M. (Ed.) *Feminism and Bioethics: Beyond Reproduction.* NY, Oxford University Press.

Women's Health Queensland Wide (1997) Information Paper on Cosmetic Surgery. Brisbane, WHQW.

Wooley, O. W. (1994) ...And Man Created 'Woman': Representations of Women's Bodies in Western Culture. In P. Fallon, M. A. Katzman and S. C. Wooley (Eds) *Feminist Perspectives on Eating Disorders.* NY, The Guildford Press.

Wyke, M. (Ed.) (1998) *Gender and the Body in the Ancient Mediterranean.* Oxford, Blackwell.

Yalom, M. (1997) *A History of the Breast.* London, Harper Collins.

Yardley, L. (1997) Essay Review: Reconstructing the Body Concept. *British Journal of Psychology*, 88, 709–13.

Young, I. M. (1990) *Throwing Like a Girl and Other Essays in Feminist Philosophy and Social Theory.* Bloomington, Indiana University Press.

Young, I. M. (1998) Breasted Experience: The Look and the Feeling. In R. Weitz (Ed.) *The Politics of Women's Bodies: Sexuality, Appearance and Behaviour.* NY, Open University Press.

Zola, I. K. (1972) Medicine as an Institution of Social Control. *Sociological Review*, 20, 487–504.

Zones, J. S. (1997) Beauty Myths and Realities and their Impact on Women's Health. In S. B. Ruzek, V. L. Olesen and A. E. Clarke (Eds) *Women's Health: Complexities and Differences.* Columbus, Ohio State University Press.

Index